Digital Relays

Power system protection is a practical area that requires extensive knowledge and experience. The organized structure, succinct illustration, and detailed programming examples provided in this book will benefit all levels of readers, including graduate students who are studying courses in electric power systems, as well as engineers who are working in electric utility companies, relay vendors, and consulting firms.

- Abstruse principles and terminologies of relay functions and devices are demystified with 87 succinct illustrations.
- Each chapter is provided with a summary of key points and a reference list that precisely guides readers to pertinent publications for further details.
- Eleven representative examples with specific industry backgrounds are thoroughly illustrated. Twelve problems are provided in key chapters to facilitate readers to establish a comprehensive understanding of relay functions.

Digital Relays
Principles and Programming

Hangtian Lei and Brian K. Johnson

CRC Press
Taylor & Francis Group
Boca Raton London New York

CRC Press is an imprint of the
Taylor & Francis Group, an **informa** business

Designed cover image: Shutterstock

MATLAB® and Simulink® are trademarks of The MathWorks, Inc. and are used with permission. The MathWorks does not warrant the accuracy of the text or exercises in this book. This book's use or discussion of MATLAB® or Simulink® software or related products does not constitute endorsement or sponsorship by The MathWorks of a particular pedagogical approach or particular use of the MATLAB® and Simulink® software.

First edition published 2026
by CRC Press
2385 NW Executive Center Drive, Suite 320, Boca Raton FL 33431

and by CRC Press
4 Park Square, Milton Park, Abingdon, Oxon, OX14 4RN

CRC Press is an imprint of Taylor & Francis Group, LLC

© 2026 Hangtian Lei and Brian K. Johnson

ISBN: 978-1-041-04703-2 (hbk)
ISBN: 978-1-041-04705-6 (pbk)
ISBN: 978-1-003-62948-1 (ebk)

DOI: 10.1201/9781003629481

Typeset in Times
by Apex CoVantage, LLC

Contents

Preface

Power system protection is a practical and broad area that requires extensive knowledge and experience. Currently, most textbooks available in this area are lengthy and not friendly to readers who do not have many years of experience in practice. Digital relays are the cornerstone for power system protection. The major purpose of this book is to provide a concise and organized illustration that covers the principles of commonly used digital relay functions.

The material of this book is based on two graduate-level power system protection courses we have been offering at the University of Idaho for over eight years. Our audiences include full-time on-campus students and part-time students from industry. The course content has been constantly updated according to the feedback from academic and industry audiences. In this book, we provide approximately ten example problems with detailed solution procedures in key chapters (e.g., Chapters 2, 4, 5, 6, and 9). These examples are developed with a specific industry background, and they are representative of the key topics in power system protection, such as instrument transformers, overcurrent protection, directional supervision, distance protection, and differential protection. Besides, we have carefully selected twelve practice problems in key chapters. By referring to pertinent principles and examples illustrated in the book, readers are able to solve these problems with their own effort and establish a comprehensive understanding of digital relays.

This book includes ten chapters, and they follow a natural sequence. As a textbook for a one- or two-semester course, it is impossible to cover all the aspects of digital relays. We choose to illustrate the most important theoretical principles and practical aspects, to the extent that readers can establish a clear view of the broad area of power system protection, and obtain the necessary knowledge of relay settings, devices, functions, and logics for their industry projects. Each chapter is provided with a summary of key points and a reference list that precisely guides readers to pertinent publications for further details.

According to our observation and students' feedback, most textbooks in power system protection focus on theoretical aspects such as fault analysis and relay setting calculation, but have not included adequate practical aspects. This book not only provides theoretical illustrations of protection functions but also includes implementation aspects in actual digital relays, such as signal processing (Chapters 3, 6, and 9), phasor computation (Chapter 3), directional supervision (Chapters 5 and 6), and relay logics (Chapters 6, 7, and 9). These aspects are indispensable for industry applications. We avoid using tedious mathematical procedures or verbose illustrations in the text. Abstruse principles and terminologies of relay functions and devices are demystified with approximately 87 succinct illustrations.

A unique feature of this book is that it includes the implementation of relay functions using MATLAB programming, which is absent in other power system protection books. As a highly practical area, having the opportunity to operate and configure relay devices is essential for the study of power system protection. However, access to actual relay devices is not available in many universities and companies. The MATLAB programs included in this book provide readers with a convenient option to test and configure digital relay functions on their own computers. These programs cover the aspects of signal processing, phasor computation, distance protection, directional supervision, and differential protection in digital relays. With an organized programming structure and clear annotations provided in Chapters 3, 4, 6, and 9 and Appendix A, readers can flexibly customize the relay settings and extend the functions for their academic research and industry projects. Furthermore, the MATLAB code is compatible with Octave, an open-source software environment using nearly the same syntax as MATLAB, which is convenient for industry users in relay design and testing.

This book could be used as a textbook for graduate students and senior undergraduate students in electric power engineering. It could also be used as a reference book for engineers who are working in electric utility companies, relay vendors, and consulting firms. Before reading this book, readers need to have knowledge of power system steady-state and fault analysis, which is typically included in a senior undergraduate electric power engineering curriculum.

Hangtian Lei
Brian K. Johnson
May 2025

Authors

Hangtian Lei is an Associate Professor in the Department of Electrical and Computer Engineering at the University of Idaho, Moscow, Idaho, USA, where he has been working since 2017. He earned a BE degree in electrical engineering at Huazhong University of Science and Technology, Wuhan, China, and a PhD degree in electrical engineering at Texas A&M University, College Station, Texas, USA. His teaching and research interests include power system protection, power system reliability, and power system planning. He is a senior member of IEEE.

Brian K. Johnson is a University Distinguished Professor and the Schweitzer Engineering Laboratories (SEL) Endowed Chair in Power Engineering in the Department of Electrical and Computer Engineering at the University of Idaho, Moscow, Idaho, USA. He earned his BS, MS, and PhD degrees in electrical engineering at the University of Wisconsin–Madison, Madison, Wisconsin, USA. He has been working at the University of Idaho for over 30 years, where he was the chair of the Department of Electrical and Computer Engineering from 2006 to 2012. His teaching and research interests include power system protection, HVDC transmission, and applications of power electronics in power systems. He is active with the IEEE Power and Energy Society, where he was the chair of the Power and Energy Education Committee from 2014 to 2015 and of the IEEE HVDC and FACTS subcommittee from 2018 to 2020. He is currently a member of the editorial board for *IEEE Power and Energy Magazine*. He is a senior member of IEEE.

Abbreviations

ANSI	American National Standards Institute
CCVT	Coupling Capacitor Voltage Transformer
COMTRADE	Common Format for Transient Data Exchange
CT	Current Transformer
CTI	Coordinating Time Interval
CTR	Current Transformer Ratio
CVT	Capacitive Voltage Transformer
DCUB	Directional Comparison Unblocking
DUTT	Direct Underreaching Transfer Trip
ES	Ethernet Switch
FIDS	Fault Identification Selection
IEC	International Electrotechnical Commission
IED	Intelligent Electronic Device
IEEE	Institute of Electrical and Electronics Engineers
IPSD	Inner Power Swing Detection
LL	Line-to-Line
MTA	Maximum Torque Angle
OOST	Out-of-Step Tripping
OPSD	Outer Power Swing Detection
OSBD	Out-of-Step Blocking Delay
POTT	Permissive Overreaching Transfer Trip
PSB	Power Swing Blocking
PUTT	Permissive Underreaching Transfer Trip
RMS	Root Mean Square
RS	Rate of Sampling
SCV	Swing Center Voltage
SLG	Single-Line-to-Ground
TDS	Time Dial Setting
TOWI	Trip-on-the-Way-In
TOWO	Trip-on-the-Way-Out
VT	Voltage Transformer
VTR	Voltage Transformer Ratio

1 Introduction

Power system protection is a branch of electric power engineering that deals with the protection of power systems from faults and disturbances. It is a critical part of electric power systems that safeguards equipment and ensures the safe operation of the power grid. When a fault occurs in a power system, protective relays are expected to detect the fault and isolate the faulted apparatus (e.g., transmission lines or transformers) by sending tripping signals to the corresponding circuit breakers.

1.1 PROTECTION SYSTEM COMPONENTS

Power system protection typically involves a set of devices and equipment for measurement, signal processing, protection algorithm computation, and actuation purposes [1]. Devices and equipment include current transformers (CTs), voltage transformers (VTs), merging units, protective relays, communication networks, and circuit breakers [2, 3]. The typical architecture of a digital protection system is shown in Figure 1.1.

Current transformers (CTs) and voltage transformers (VTs) are also called instrument transformers. The current and voltage levels in a power system can reach hundreds of kilo-Amperes or Volts. The main purpose of using instrument transformers is to scale down such a high current or voltage measured from the power system to a much smaller value that can be processed by electronic devices. A more detailed illustration of instrument transformers will be provided in Chapter 2 of this book.

After currents or voltages are scaled down, they are digitized by merging units and transmitted to digital protective relays through communication links. Protective relays perform protection functions based on input currents and/or voltages. If a fault is determined, tripping signals will be sent to the circuit breakers. A general flow chart with protection system components is shown in Figure 1.2. It should be noted that there could be variations in practical fields. In some applications, the digitalization process is completed by modules inside protective relays.

1.2 PROTECTIVE RELAYS

Protective relays are the core components for power system protection. In the past, electromechanical relays with standalone functions were used for electric power systems. An example of electromechanical relay is shown in Figure 1.3. It is an overcurrent relay made by General Electric. An electromechanical overcurrent relay includes a current coil that is placed in series with the protected circuit. When current flows through the coil, a magnetic field is produced, which generates a magnetic force on a moving element (e.g., a disk). As the current increases,

DOI: 10.1201/9781003629481-1

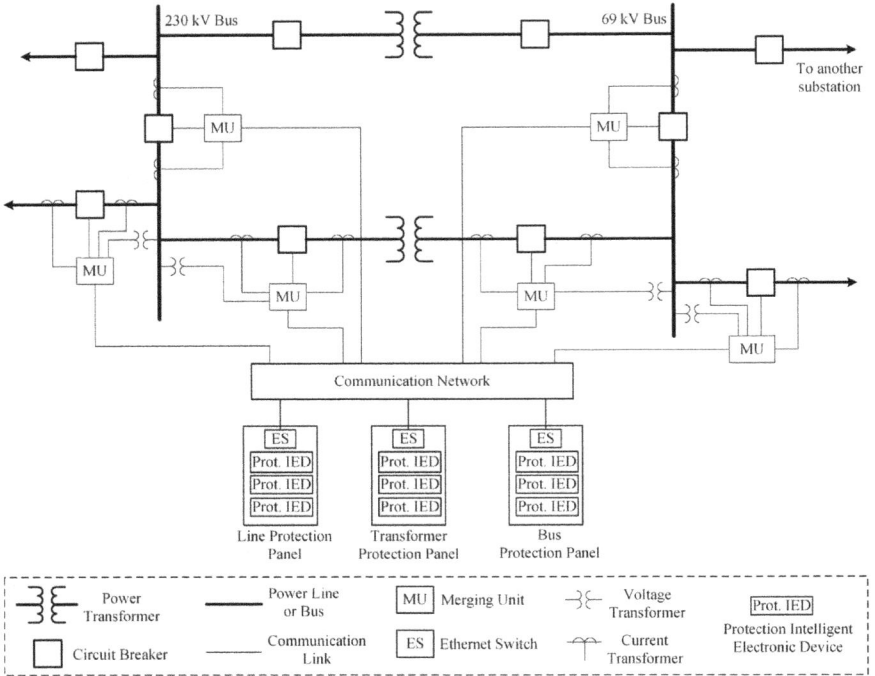

FIGURE 1.1   Typical architecture of a digital protection system.

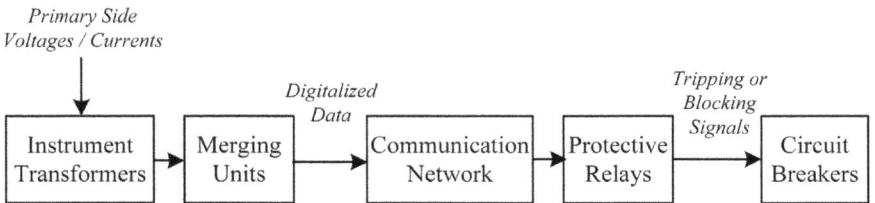

FIGURE 1.2   General flow chart for a protection system.

the magnetic force also increases. If the magnetic force exceeds the restraining force from a spring, the moving element will move. The position of relay contact(s) will be changed by the movement of the moving element. Depending on the specific design of control circuit, the position change of relay contact(s) could close or open the circuit or trigger other actions.

Electromechanical relays typically offer basic protection functions and are not suitable for protection schemes with complex logics or communication. The moving parts of electromechanical relays are susceptible to wear and tear, which require periodic calibration, testing, or replacement.

FIGURE 1.3   Example of electromechanical relay. (Courtesy of General Electric.)

With the advancement of technologies, electromechanical relays have been mostly replaced by multifunctional, communicative, and programmable digital relays [4].

Digital relays, also known as *microprocessor-based relays* or *protection intelligent electronic devices (IEDs)*, are electronic devices that constantly take voltage or current measurements as inputs and perform protection functions. If a fault is determined by one or multiple protection functions, the relay will send tripping commands to circuit breakers and communication signals to other devices to isolate the faulted section(s) from the rest of the power system. Examples of digital relays are shown in Figure 1.4. The left one is a transmission line protection

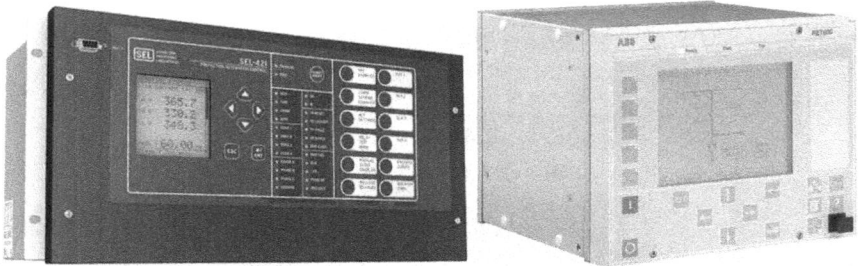

FIGURE 1.4   Examples of digital relays. (Courtesy of SEL and ABB.)

relay made by Schweitzer Engineering Laboratories, Inc. (SEL). The right one is a transformer protection relay made by ABB.

Because of the electronic control and lack of mechanical wear, digital relays typically offer greater accuracy and consistent performance compared to electromechanical relays. Digital relays also provide advanced features such as event recording and remote monitoring.

1.3   PROTECTION FUNCTIONS

Protection functions are programmed inside digital relays. Some functions use current measurements only, such as overcurrent protection and current differential protection. Some functions utilize both current and voltage measurements, such as distance protection. In practical fields, ANSI/IEEE device numbers are used to denote protection functions and devices [1]. Some of the commonly used ANSI/IEEE device numbers are listed in Table 1.1.

Digital relays are multifunctional. For example, a digital relay for transmission line protection typically includes distance, directional overcurrent, and synchronism-check functions. A digital relay for transformer protection typically includes differential, undervoltage, overvoltage, and synchronism-check functions.

1.4   ORGANIZATION OF THIS BOOK

In this book, we mainly illustrate the principles of commonly used protection functions in digital relays. The remainder of this book is organized as follows. Chapter 2 illustrates the principles of current and voltage transformers. Chapter 3 introduces the modeling and programming of digital relay functions using MATLAB. Chapter 4 illustrates overcurrent protection. Chapter 5 illustrates directional supervision. Chapter 6 illustrates distance protection. Chapter 7 illustrates communication-aided protection schemes. Chapter 8 illustrates power swing blocking and out-of-step tripping. Chapter 9 illustrates differential protection. Chapter 10 illustrates time-domain protection.

TABLE 1.1
Commonly Used ANSI/IEEE Device Numbers

Number	Device or Function	Number	Device or Function
1	Master Element	50	Instantaneous Overcurrent Protection
2	Time-Delay Starting or Closing Relay	51	AC Inverse Time Overcurrent Protection
7	Rate of Change Relay	52	AC Circuit Breaker
7F	Alternative Number for Rate of Change of Frequency Relay	59	Overvoltage Relay
21	Distance Protection	67	AC Directional Overcurrent Protection
21G	Ground Distance	68	Power Swing Blocking
21P	Phase Distance	78	Phase Angle Measuring or Out-of-Step Tripping
TD21	Incremental Quantity-Based Distance	79	AC Reclosing Relay
24	Volts per Hertz Relay	81	Frequency Relay
25	Synchronizing or Synchronism-Check	85	Pilot Communications, Carrier, or Pilot-Wire Relay
27	Undervoltage Protection	87	Differential Protection
32	Directional Function	87B	Bus Differential
37	Undercurrent or Underpower Protection	87T	Transformer Differential
40	Field (Over/Under Excitation) Relay	TW87	Traveling Wave-Based Differential

BIBLIOGRAPHY

[1] J. L. Blackburn and T. J. Domin, *Protective Relaying: Principles and Applications*, 3rd Ed. CRC Press, 2006.

[2] A. F. Sleva, *Protective Relay Principles*. CRC Press, 2009, pp. 112–113.

[3] H. Lei, C. Singh, and A. Sprintson, "Reliability modeling and analysis of IEC 61850 based substation protection systems," *IEEE Transactions on Smart Grid*, vol. 5, no. 5, pp. 2194–2202, September 2014.

[4] P. M. Anderson, C. Henville, R. Rifaat, B. Johnson, and S. Meliopoulos, *Power System Protection*, 2nd Ed. Wiley, 2022.

2 Instrument Transformers

As briefly mentioned in Chapter 1, instrument transformers mainly include current transformers (CTs) and voltage transformers (VTs). The current and voltage levels in a power system could reach hundreds of kilo-Amperes or Volts. The main purpose of using instrument transformers is to scale down such a high current or voltage measured from the power system to a much smaller value that can be processed by electronic devices. In this chapter, we will illustrate the principles of instrument transformers and their equivalent circuits for power system protection studies.

2.1 CURRENT TRANSFORMERS

The typical construction and principle of a current transformer (CT) are shown in Figure 2.1. In the construction diagram (left), the primary side conductor is indicated by number 5, and secondary windings are indicated by number 11. The primary side is a power transmission or distribution line that carries AC current I_P. The secondary side current I_S flows in the CT secondary circuit. Ideally, the current ratio equals the inverse of the turn ratio, as shown in Equation (2.1). The current ratio is also called current transformer ratio (CTR). In industry, the secondary side current is typically rated at 5 A (RMS). The primary side turn number $N_{1CT} = 1$. The secondary side turn number N_{2CT} depends on the primary side current rating.

$$CTR = \frac{I_P}{I_S} = \frac{N_{2CT}}{N_{1CT}} \tag{2.1}$$

2.2 CT TRANSIENTS

Practically, CT secondary current I_S may have distortion due to core saturation [1, 2]. A simplified CT equivalent circuit is typically used to study the transient features of CTs, as shown in Figure 2.2. A current source (I'_P) is used to represent the scaled-down current [3]. This current is calculated directly from the turn ratio and is called the ratio current. A current source is used here because the change of CT secondary circuit has a minor effect on the currents in the primary side (i.e., power lines). The core of the CT is represented as an equivalent inductance L_M. The actual CT secondary current is represented as I_S. R_B and L_B represent the equivalent impedance of secondary circuit and they are called CT secondary burden or relay burden.

When the power system is operating in a normal condition, the core has almost no saturation and the secondary current is almost the same as the ratio current.

DOI: 10.1201/9781003629481-2

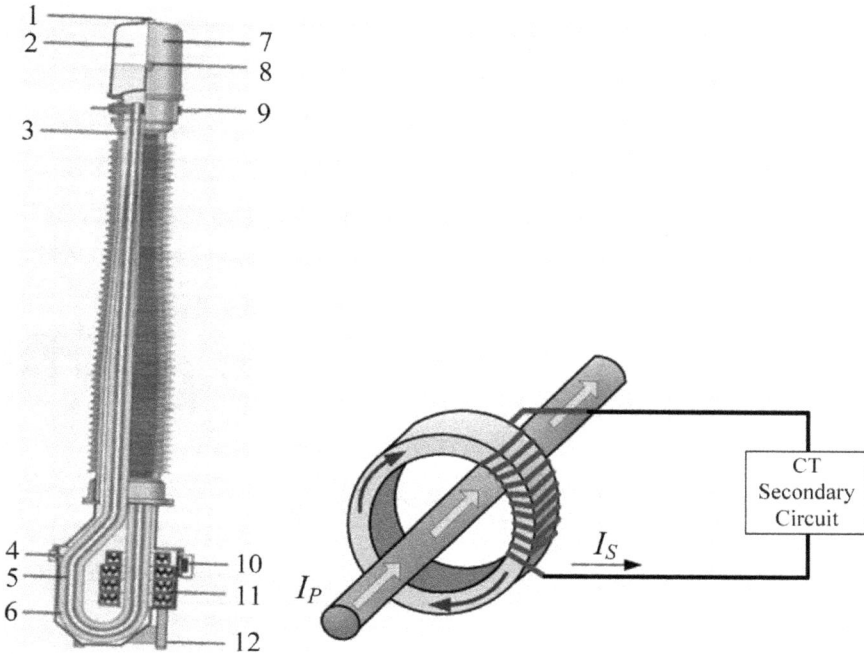

FIGURE 2.1 Construction and principle of a current transformer.

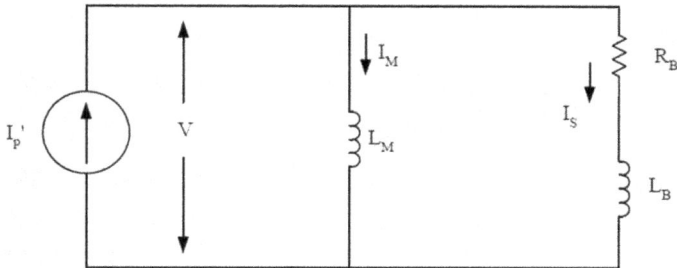

FIGURE 2.2 Simplified CT equivalent circuit.

When a fault occurs, the difference between the secondary current and ratio current could be significant, especially in the first several cycles after fault occurrence [4]. Figure 2.3 shows an example comparing a CT secondary current with a ratio current, in which Figure 2.3 (a) shows the unfiltered currents and Figure 2.3 (b) shows the currents after passing through a filter. I_{CT_SEC} represents the CT secondary current and I_{RATIO} represents the ratio current. The distortion affects both magnitude and angle, as shown in Figure 2.4. In Figure 2.4 (b), the angle of I_{RATIO} is used as a reference, and the plotted curve is the angle of I_{CT_SEC}.

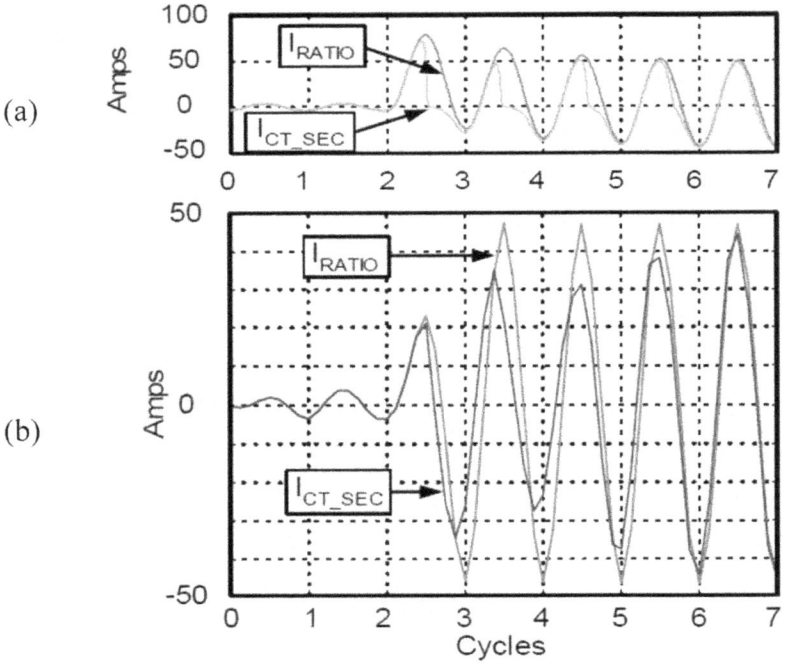

FIGURE 2.3 Comparison between a CT secondary current and a ratio current: (a) unfiltered and (b) filtered.

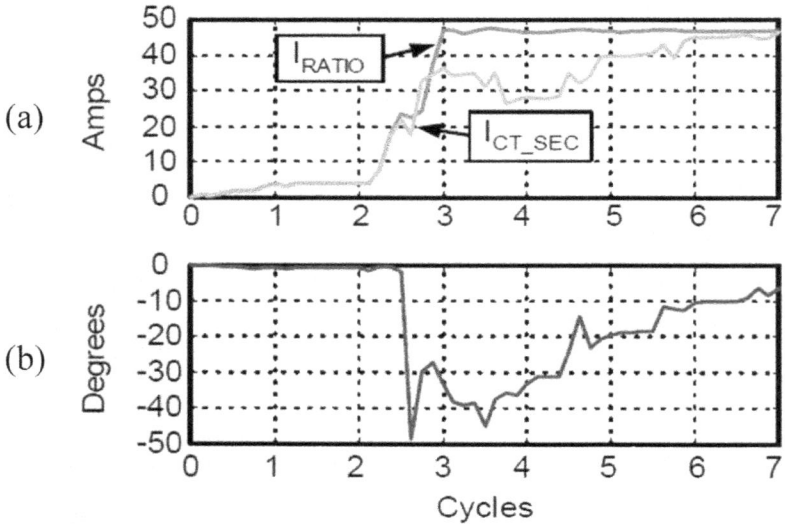

FIGURE 2.4 (a) Current magnitudes and (b) current angle.

2.3   C-CLASS CTS

The accuracy of a CT refers to the ability to reproduce the primary current in secondary amperes in both wave shape and magnitude. C-Class CTs are a type of CT widely used in power system protection industry. ANSI/IEEE C57.13 standard requires that when a C-Class CT is connected to a standard burden and the secondary current reaches 20 times its rated value, the measurement error should not exceed 10% [1, 2].

C-Class CTs are named according to its knee voltage, such as C100, C200, C600, and C800. The knee voltage refers to the RMS value of the voltage V in Figure 2.2 when the CT secondary current reaches 20 times its rated value under the condition that the CT is connected to a standard burden. The CT secondary current is typically rated at 5 A. For a C600 CT, its standard burden can be calculated as 600 V / (5 A * 20) = 6 Ω.

Figure 2.5 shows the excitation curve of a C600 CT. The vertical axis refers to the RMS value of excitation voltage, which is the voltage V shown in Figure 2.2. The horizontal axis refers to the RMS value of the current I_M shown in Figure 2.2.

It can be seen from Figure 2.5 that when the excitation voltage is below the knee voltage, the excitation current is negligible, which means the difference

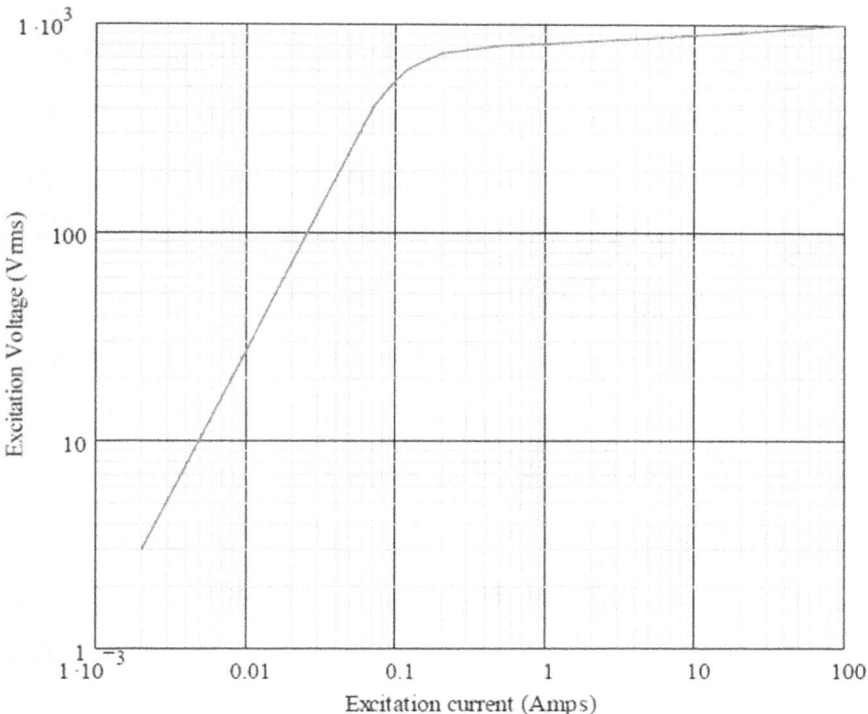

FIGURE 2.5   Excitation curve of a C600 CT.

between the ratio current and actual CT secondary current is minor. When the excitation voltage exceeds the knee voltage, the difference between the ratio current and the actual CT secondary current becomes more significant.

In engineering practice, sometimes we need to select the C-Class rating of a CT based on system operation. If the condition shown in Equation (2.2) is satisfied, a C-class CT will generally not saturate.

$$\left(1+\frac{X}{R}\right)\frac{I_{mag}}{I_{rated}}\frac{Z_B}{\left(\frac{V_{knee}}{100}\right)} < 20 \tag{2.2}$$

The X/R is the reactance-to-resistance ratio of the system. Considering the X/R ratio means that the worst-case DC component in the fault current is taken into account. The I_{mag} is the actual current magnitude and I_{rated} is the magnitude of rated current. Both I_{mag} and I_{rated} should use quantities on the same side (primary or secondary). Z_B is the relay burden and V_{knee} is the knee voltage of the CT.

Example 2.1

You need to determine the C-class rating for 2000/5 CTs applied to a transmission system. The X/R ratio for the fault impedance in the worst-case fault seen by the CTs is 8. The relay burden is 6.25 Ω.

(a) If the decaying DC offset is neglected and the maximum fault current is 21000 A on the primary side, determine the C-class rating (knee voltage) so that the CT will not saturate.
(b) If the decaying DC offset is included and the maximum fault current is 21000 A on the primary side, determine the C-class rating (knee voltage) so that the CT will not saturate.
(c) Repeat parts (a) and (b) for a fault current of 27000 A with a relay burden of 3.75 Ω and the same X/R ratio.

Solution:

(a) CTR $= 2000/5 = 400$
The maximum fault current on the secondary side $I_{fmax} = 21000/\text{CTR} = 52.5$ A .
The rated current magnitude on the secondary side $I_{rated} = 5$ A.
Now, to find the knee voltage if the DC offset is neglected:
$V_{knee\_withoutDC} = I_{fmax} * Z_B = 52.5 \text{ A} * 6.25 \text{ Ω} = 328.125$ V
Selecting C-class rating of 328.125 V is needed. In reality, round up to C400.
(b) Now, to find the knee voltage if the DC offset is considered:
$V_{knee\_withDC} = I_{fmax} * Z_B * (1 + X/R) = 52.5 \text{ A} * 6.25 \text{ Ω} * (1+8) = 2953.125$ V

Selecting C-class rating of 2953.125 V is needed. In reality, round up to C3000.

(c) $CTR = 2000/5 = 400$

The maximum fault current on the secondary side $I_{fmax} = 27000/CTR = 67.5$ A.

The rated current magnitude on the secondary side $I_{rated} = 5$ A.

Now, to find the knee voltage if the DC offset is neglected:

$$V_{knee\_withoutDC} = I_{fmax} * Z_B = 67.5 \text{ A} * 3.75 \ \Omega = 253.125 \text{ V}$$

Selecting C-class rating of 253.125 V is needed. In reality, round up to C300.

Now to find the knee voltage if the DC offset is considered:

$$V_{knee\_withDC} = I_{fmax} * Z_B * (1 + X/R) = 67.5 \text{ A} * 3.75 \ \Omega * (1+8) = 2278.125 \text{ V}$$

Selecting C-class rating of 2278.125 V is needed. In reality, round up to C2300.

2.4 VOLTAGE TRANSFORMERS

Voltage transformers (VTs) are also called potential transformers (PTs). The main function of a VT is to scale down a high voltage acquired from the power system to a much smaller value that can be processed by electronic devices. There are three main types of VTs: wire wound, optical, and capacitive. The capacitive type of VTs is called capacitive voltage transformers (CVTs), or coupling capacitor voltage transformers (CCVTs). A CVT is a CCVT without carrier accessories. In industry, the terms "CVT" and "CCVT" are typically used interchangeably. In this section, we will mainly illustrate the principles of CVTs (or CCVTs).

The general structure of a CVT [5] is shown in Figure 2.6. The line voltage refers to the voltage acquired at a measurement point from a power line. The compensating reactor is used to compensate for the voltage phase angle shift caused

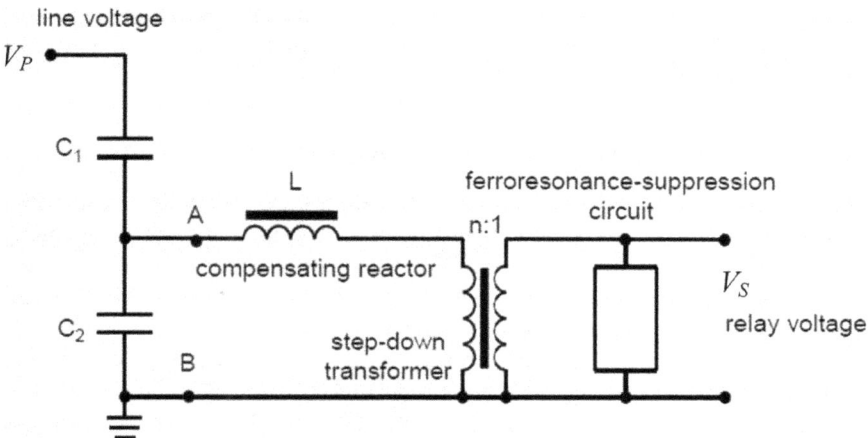

FIGURE 2.6 CVT general structure.

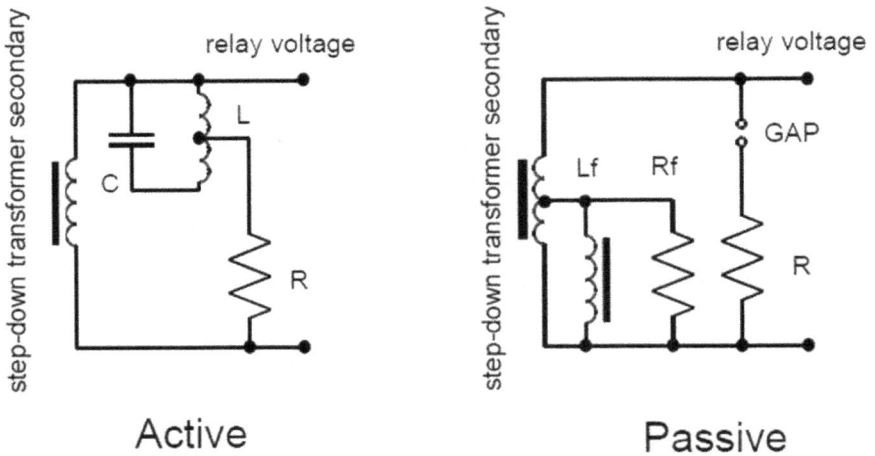

FIGURE 2.7   Active and passive ferroresonance suppression circuits.

by capacitors C_1 and C_2. The net voltage transformer ratio (VTR) is calculated using Equation (2.3). It should be noted that the net VTR includes the overall effect of C_1, C_2, the compensating reactor, and the step-down transformer.

$$VTR = \frac{V_P}{V_S} = \frac{N_{1VT}}{N_{2VT}} \tag{2.3}$$

Ferroresonance can cause overvoltage and can lead to overheating or damage. A ferroresonance suppression circuit is used to limit or prevent ferroresonance. A ferroresonance suppression circuit can be passive or active. A passive circuit typically uses a damping resistance and a saturable inductor, while an active circuit typically uses a parallel-series RLC filter. Examples of active and passive ferroresonance suppression circuits are shown in Figure 2.7.

2.5   TRANSIENTS SIMULATION OF A CVT

Sometimes it is necessary to capture the transient features of CVTs in relay functional testing [6, 7], especially in the testing of time-domain protection functions (e.g., traveling-wave-based protection functions). In this section, we build the simulation model of a CVT using an electromagnetic transients simulation program and test its response in steady-state and transient conditions.

The equivalent circuit of a CVT is shown in Figure 2.8. The V_{PRI} refers to the primary side voltage, and V_{SEC} refers to the secondary side voltage. Parameters of the circuit are provided in Table 2.1. The net VTR is 2000. The base frequency is set as 60 Hz in the simulation.

The simulation results under a normal condition and a fault condition are shown in Figures 2.9 and 2.10, respectively. In this simulation, the secondary

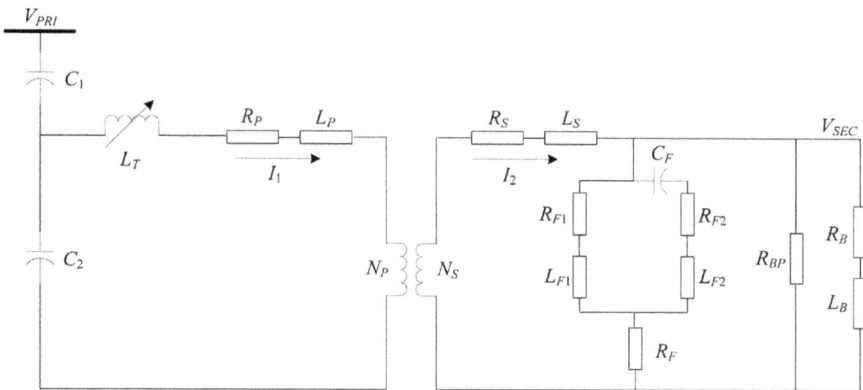

FIGURE 2.8 CVT equivalent circuit.

TABLE 2.1
Parameters of CVT Equivalent Circuit

Parameter Name	Value	Unit
C_1	0.012896	uF
C_2	0.26397	uF
L_T	20.953	H
R_P	474.0	Ω
L_P	4.46	H
N_P	11,000	turns
R_S	0.18	Ω
L_S	0.00047	H
N_S	115	turns
R_B	400.9	Ω
L_B	1.84	H
R_{BP}	2,298.0	Ω
R_{F1}	1.06	Ω
L_{F1}	0.01	H
C_F	8.0	uF
R_{F2}	4.24	Ω
L_{F2}	0.394	H
R_F	40	Ω

voltage distortion is minor even under a fault condition. The location of fault and the time instant of fault occurrence may affect the CVT response. In practical fields, scenarios with various possible fault locations and fault occurrence instants should be tested.

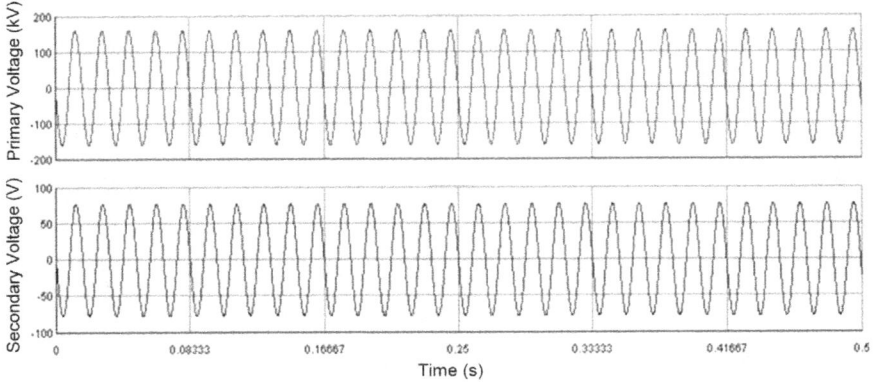

FIGURE 2.9 CVT response under a normal condition.

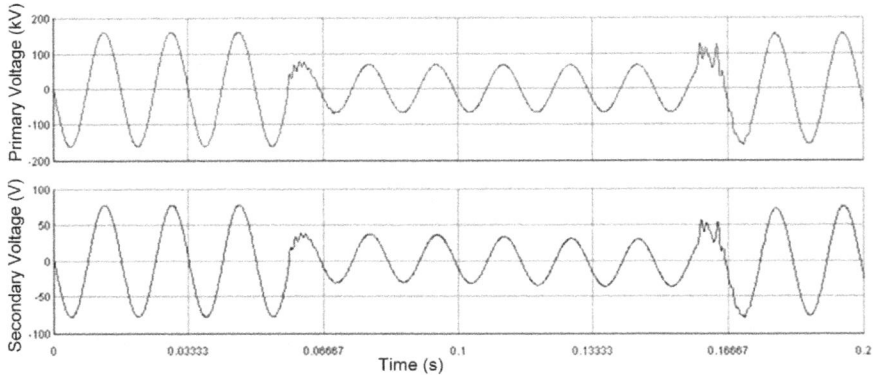

FIGURE 2.10 CVT response under a fault condition.

2.6 IMPEDANCE CONVERSION

If we know a primary side impedance Z_P, how can we represent the equivalent impedance on the secondary side (Z_S) in terms of CT ratio (CTR) and VT ratio (VTR)?

We know from Equations (2.1) and (2.3) that:

$$CTR = \frac{I_P}{I_S} = \frac{N_{2CT}}{N_{1CT}} \text{ and } VTR = \frac{V_P}{V_S} = \frac{N_{1VT}}{N_{2VT}}$$

Therefore, the secondary side equivalent impedance can be obtained using Equation 2.4:

$$Z_S = \frac{V_S}{I_S} = \frac{V_P \left(N_{2VT} / N_{1VT} \right)}{I_P \left(N_{1CT} / N_{2CT} \right)} = \frac{V_P \left(1 / VTR \right)}{I_P \left(1 / CTR \right)} = Z_P \frac{CTR}{VTR}. \tag{2.4}$$

2.7 SUMMARY

Instrument transformers mainly include current transformers and voltage transformers. Their principles, typical types, equivalent circuits, steady-state features, and transient features have been illustrated in this chapter.

2.8 PROBLEMS

Problem 2.1

A CT with the excitation curve described in Table 2.2 is carrying 1128 A (RMS) primary side current and a resistive burden of 4 Ω. The CTR = 1200/5. Neglecting the core loss resistance of the CT.

(1) Calculate the approximate initial voltage that would result across the CT secondary winding if the CT secondary winding is accidentally opened. Note: This is an example for theoretical analysis only. In practical fields, opening a CT secondary winding is not allowed. If the CT secondary winding is accidentally opened, the CT would probably fail before reaching the initial overvoltage.

(2) Calculate the approximate final voltage it would reach if the insulation survives the initial overvoltage. Explain where the current is flowing.

Problem 2.2

The simplified circuit of a coupling capacitor voltage transformer (CCVT, also known as CVT) is shown in Figure 2.11. The parameter values are listed in Table 2.3.

(1) Determine a Laplace domain transfer function for the relay voltage (the voltage across the load resistance R_0) in response to a change in the input voltage.

(2) Plot the frequency response (using a Bode plot) of the magnitude of the output voltage from 20 Hz to the 25th harmonic of 60 Hz.

(3) Plot the responses for the output voltage versus time with the two sets of CCVT parameters listed below using a circuit simulation program when the primary voltage goes to zero due to a fault occurring at a voltage peak with an ideal source.

TABLE 2.2
Table for Problem 2.1

Excitation Current (A)	0.001	0.04	0.1	0.12	0.14	0.2	0.3	0.4	40
Excitation Voltage (V)	0.09	90	428	520	600	700	780	800	927

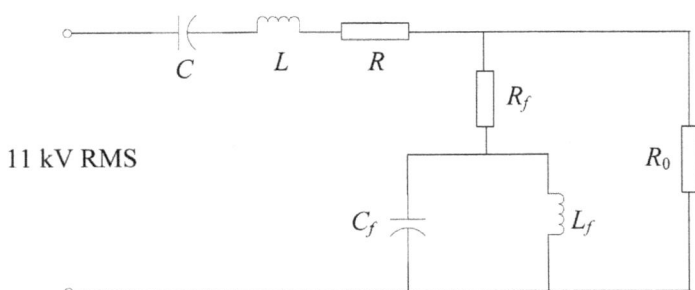

FIGURE 2.11   Circuit diagram for Problem 2.2.

TABLE 2.3
Parameters of Two CCVTs

	Medium Capacitance (Energy) CCVT	High Capacitance (Energy) CCVT	Unit
R_0	$1.03997*10^5$	$2.08584*10^5$	Ω
L_f	315.3	616.35	H
C_f	$0.0285*10^{-6}$	$0.01134*10^{-6}$	F
R_f	77379	148519	Ω
R	3289	1536	Ω
C	$9.1605*10^{-8}$	$0.162442*10^{-6}$	F
L	76.136	48.136	H

BIBLIOGRAPHY

[1] P. M. Anderson, C. Henville, R. Rifaat, B. Johnson, and S. Meliopoulos, *Power System Protection*, 2nd Ed. Wiley, 2022.

[2] J. C. Das, *Power System Protective Relaying*. CRC Press, 2017.

[3] R. Garrett, W. C. Kotheimer, and S. E. Zocholl, "Computer simulation of current transformers and relays for performance analysis," in 14th Annual Western Protective Relay Conference, pp. 20–23, 1987.

[4] R. Folkers, *Determine Current Transformer Suitability Using EMTP Models*. Schweitzer Engineering Laboratories, 1999. https://cdn.selinc.com/assets/Literature/Publications/Technical%20Papers/6096_DetermineCurrent_991007_Web.pdf?v=20150812-085604.

[5] D. Hou and J. Roberts, "Capacitive voltage transformer: transient overreach concerns and solutions for distance relaying," in 1996 Canadian Conference on Electrical and Computer Engineering, vol. 1, pp. 119–125, 1996, DOI: 10.1109/CCECE.1996.548052.

[6] A. Sweetana, "Transient response characteristics of capacitive potential devices," *IEEE Transactions on Power Apparatus and Systems*, vol. PAS-90, no. 5, pp. 1989–2001, 1971.

[7] B. Kasztenny, D. Sharples, V. Asaro, and M. Pozzuoli, "Distance relays and capacitive voltage transformers-balancing speed and transient overreach," in 53rd Annual Conference for Protective Relay Engineers, pp. 11–13, 2000.

3 Modeling Relay Functions Using MATLAB

MATLAB is a programming language widely used in academia and industry. Its user-friendly environment and extensive libraries make it convenient for technical programming in various science and engineering fields [1]. In this chapter, we illustrate the overall structure and coding of a MATLAB program for the modeling and simulation of relay functions. This chapter mainly illustrates the modules for input data reading and signal processing. The programming details for protection functions and relay logics will be covered in later chapters. The MATLAB programming code can be conveniently implemented on readers' own computers, which will greatly enhance readers' understanding of relay functions and logics.

3.1 OVERALL STRUCTURE

A digital relay typically consists of modules with the following functions:

(1) Input data reading.
(2) Input data processing.
(3) Protection functions and logics.
(4) Outputting the results.

We will use a distance relay as an example to illustrate detailed functions in a digital relay model. Each module is saved as a MATLAB script file (.m). All the MATLAB scripts and input data files are put in one folder, as shown in Figure 3.1.

The *Relay1_main.m* is the master file. Running this file will run the entire relay program. The MATLAB code of the *Relay1_main.m* file is shown as follows:

```
%Relay1_main.m, the master file of a distance relay
%Begin
clear all;
close all;
clc;
%Read voltage and current data
Relay1_readdata;
%Configure the Relay1 settings
Relay1_setting;
%Process the input currents and voltages through a filter
Relay1_filter;
```

DOI: 10.1201/9781003629481-3

Name

Relay1_main.m

Relay1_readdata.m

Relay1_plot.m

Relay1_setting.m

Relay1_phasor_vpol.m

Relay1_21G_calc.m

Relay1_21P_calc.m

Relay1_filter.m

Relay1_21P_zone1.m

Relay1_21G_zone1.m

Relay1_directional.m

FSLG75.cfg

FSLG75.dat

FSLG75.hdr

testdata.txt

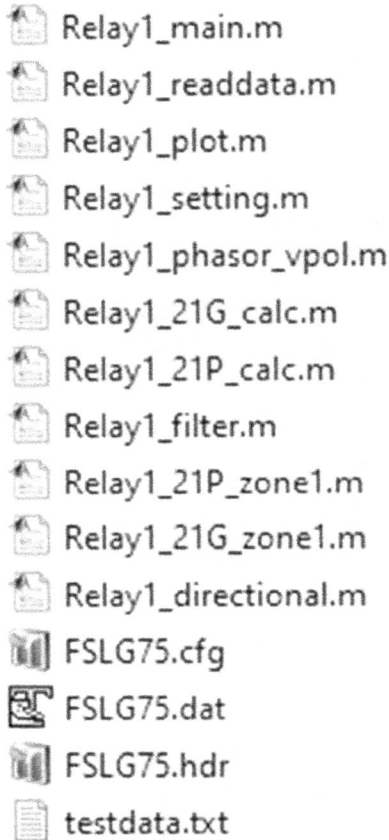

FIGURE 3.1 MATLAB script and input data files for a distance relay model.

```
%Create Phasors and Vpol
Relay1_phasor_vpol;
%Perform calculation for Ground distance elements (21G)
Relay1_21G_calc;
%Perform calculation for Phase distance elements (21P)
Relay1_21P_calc;
%Run directional elements
Relay1_directional;
%Run the pick-up logic for Zone 1 Ground distance elements
(21G)
Relay1_21G_zone1;
%Run the pick-up logic for Zone 1 Phase distance elements
(21P)
Relay1_21P_zone1;
```

```
%Output the results
Relay1_plot;
%End
```

The *Relay1_readdata.m* reads the input voltages and currents.

The *Relay1_setting.m* configures the relay settings.

The *Relay1_filter.m* performs a filtering algorithm on voltages and currents.

The *Relay1_phasor_vpol.m* creates phasors and calculates polarizing voltages for further protection functions.

The *Relay1_21G_calc.m*, *Relay1_21P_calc.m*, *Relay1_directional.m*, *Relay1_21G_zone1.m*, and *Relay1_21P_zone1.m* are the protection functions and logics for a distance relay. Their details will be illustrated in later chapters.

The *Relay1_plot.m* plots the results.

In this chapter, we will mainly illustrate the details of the *Relay1_readdata.m*, *Relay1_setting.m*, *Relay1_filter.m*, and *Relay1_phasor_vpol.m* modules.

3.2   INPUT DATA READING

Input data are read using the *Relay1_readdata.m* module. Input data include three-phase voltages and currents, which can be saved as either a text file or COMTRADE file.

3.2.1   TEXT FILE

An example of text file is shown in Figure 3.2. In this example, the text file consists of 7 columns of data, representing ground current (IR), phase A current (IA), phase B current (IB), phase C current (IC), phase A voltage (VA), phase B voltage (VB), and phase C voltage (VC), respectively. It should be noted that in this example, the sampling frequency is already known to the relay. Therefore, a column of time instant data is not included in the input file. If the sampling frequency is not known to the relay, then an additional column of time instant data should be provided, based on which the relay could calculate the corresponding frequency.

Reading a text file is simple. An example of MATLAB code is shown below. After running the code, the input data are saved as 7 vectors, IR, IA, IB, IC, VA, VB, and VC, in the MATLAB program.

```
%Reading input data from a text file
%Begin
AA = importdata('testdata.txt');
CTR = 240;
PTR = 1000;
IR = AA(:,1)/CTR;
IA = AA(:,2)/CTR;
IB = AA(:,3)/CTR;
```

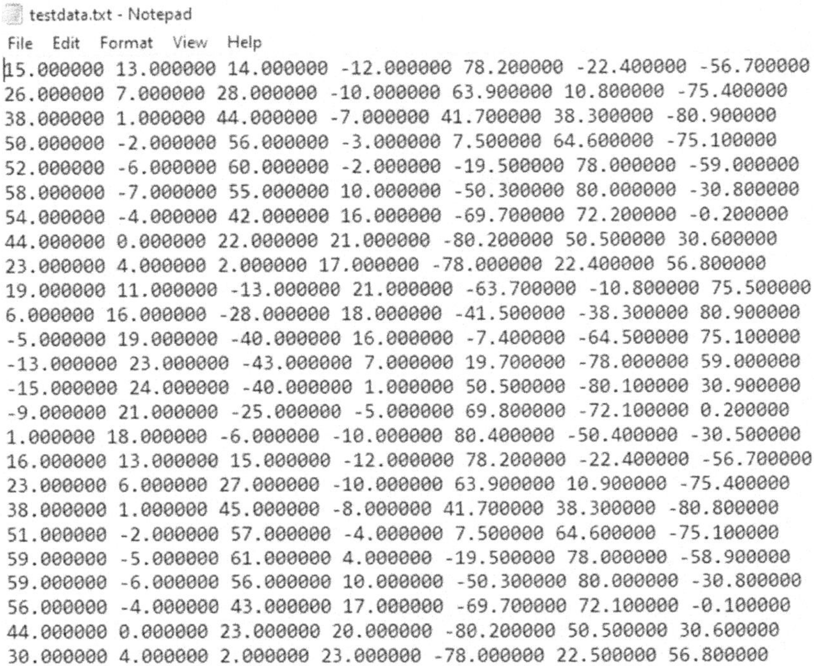

```
testdata.txt - Notepad
File  Edit  Format  View  Help
15.000000 13.000000 14.000000 -12.000000 78.200000 -22.400000 -56.700000
26.000000 7.000000 28.000000 -10.000000 63.900000 10.800000 -75.400000
38.000000 1.000000 44.000000 -7.000000 41.700000 38.300000 -80.900000
50.000000 -2.000000 56.000000 -3.000000 7.500000 64.600000 -75.100000
52.000000 -6.000000 60.000000 -2.000000 -19.500000 78.000000 -59.000000
58.000000 -7.000000 55.000000 10.000000 -50.300000 80.000000 -30.800000
54.000000 -4.000000 42.000000 16.000000 -69.700000 72.200000 -0.200000
44.000000 0.000000 22.000000 21.000000 -80.200000 50.500000 30.600000
23.000000 4.000000 2.000000 17.000000 -78.000000 22.400000 56.800000
19.000000 11.000000 -13.000000 21.000000 -63.700000 -10.800000 75.500000
6.000000 16.000000 -28.000000 18.000000 -41.500000 -38.300000 80.900000
-5.000000 19.000000 -40.000000 16.000000 -7.400000 -64.500000 75.100000
-13.000000 23.000000 -43.000000 7.000000 19.700000 -78.000000 59.000000
-15.000000 24.000000 -40.000000 1.000000 50.500000 -80.100000 30.900000
-9.000000 21.000000 -25.000000 -5.000000 69.800000 -72.100000 0.200000
1.000000 18.000000 -6.000000 -10.000000 80.400000 -50.400000 -30.500000
16.000000 13.000000 15.000000 -12.000000 78.200000 -22.400000 -56.700000
23.000000 6.000000 27.000000 -10.000000 63.900000 10.900000 -75.400000
38.000000 1.000000 45.000000 -8.000000 41.700000 38.300000 -80.800000
51.000000 -2.000000 57.000000 -4.000000 7.500000 64.600000 -75.100000
59.000000 -5.000000 61.000000 4.000000 -19.500000 78.000000 -58.900000
59.000000 -6.000000 56.000000 10.000000 -50.300000 80.000000 -30.800000
56.000000 -4.000000 43.000000 17.000000 -69.700000 72.100000 -0.100000
44.000000 0.000000 23.000000 20.000000 -80.200000 50.500000 30.600000
30.000000 4.000000 2.000000 23.000000 -78.000000 22.500000 56.800000
```

FIGURE 3.2   An example of text file for relay input.

```
IC = AA(:,4)/CTR;
VA = AA(:,5)*1000/PTR;
VB = AA(:,6)*1000/PTR;
VC = AA(:,7)*1000/PTR;
len = length(IR);
%End
```

3.2.2   COMTRADE FILES

COMTRADE (Common Format for Transient Data Exchange) is a file format
for storing voltage and current waveforms related to power system events. The
COMTRADE file format was standardized as C37.111 by the Power System
Relaying and Controls Committee (PSRC) of the IEEE Power and Energy
Society (PES).

The initial version of COMTRADE file specification was published in 1991. It
specifies that COMTRADE file format includes three file types, with the exten-
sions ".cfg", ".dat", and ".hdr", respectively. The configuration (.cfg) file specifies
signal properties for the data (.dat) file, such as the number of signals, signal
names, scaling factors, offset factors, and the number of samples. The data
(.dat) file includes digitized data values in an ASCII (American Standard Code
for Information Interchange) text format. The .hdr file is a file for displaying

```
FDLG75.cfg - Notepad
File  Edit  Format  View  Help
Analyzer,0,1999
7,7A,0D
1,IR1,,,A,2.1680E-04,8.5289E-01,0.0000E00,-32765,32765,1,1,P
2,IA1,,,A,3.5439E-05,-4.5821E-02,0.0000E00,-32765,32765,1,1,P
3,IB1,,,A,3.4863E-04,-2.3701E00,0.0000E00,-32765,32765,1,1,P
4,IC1,,,A,3.7861E-04,3.0837E00,0.0000E00,-32765,32765,1,1,P
5,V1A,,,V,5.4411E-03,-6.7022E00,0.0000E00,-32765,32765,1,1,P
6,V1B,,,V,5.2590E-03,9.2010E-01,0.0000E00,-32765,32765,1,1,P
7,V1C,,,V,5.2157E-03,-2.1350E-02,0.0000E00,-32765,32765,1,1,P
60
1
960,192
12/19/2006,01:50:49.494949
12/19/2006,01:50:49.494949
ASCII
1
```

FIGURE 3.3 An example of configuration (.cfg) file.

waveforms in a program. Only the configuration (.cfg) and data (.dat) files are mandatory. The most widely used COMTRADE version is C37.111-1999 [2]. This version added a 16-bit binary data (.dat) file format.

An example of configuration (.cfg) file is shown in Figure 3.3. The first row shows the location of data acquisition and the COMTRADE file version (1999). In this example, the data are acquired from a simulation case using a program called ATP-Analyzer. If the data are acquired from an actual relay from the field, the first row may show a substation name. The second row shows the number of signals. In this example, there are 7 signals, including 4 currents and 3 voltages. The next 7 rows include the properties of each signal. Each row includes the signal name, unit, scaling factor, offset factor, minimum or maximum values, and other control variables.

An example of data (.dat) file is shown in Figure 3.4. The first column shows the number of each row. The second column shows the time instant of each row. The next 7 columns include the uncorrected instantaneous values of the 7 signals.

For each signal shown in columns 3–9, its corrected instantaneous values should be the uncorrected values multiplied by a scaling factor and then plus an offset factor, which have been provided in the configuration (.cfg) file.

An example of MATLAB code for reading COMTRADE files is shown below. After running the code, the input data are saved as 7 vectors, IR, IA, IB, IC, VA, VB, and VC, in the MATLAB program.

```
FDLG75.dat - Notepad

File  Edit  Format  View  Help
1,0,-3934,-14086,8187,-7985,-14441,13252,2795
2,1041,-3935,-32765,8987,-6973,-32765,19933,15215
3,2083,-3934,-27894,7421,-5987,-27009,4174,25058
4,3125,-3935,-24497,6422,-5386,-23002,-6216,31397
5,4166,-3933,-13350,5043,-5158,-11416,-19541,32702
6,5208,-3936,-2832,4267,-5430,-801,-27043,29237
7,6250,-3933,10608,3598,-6070,12441,-32765,21147
8,7291,-3936,20707,3657,-7071,22040,-31583,9988
9,8333,-3933,29509,3980,-8191,30136,-27304,-2821
10,9375,-3936,32553,4926,-9349,32369,-17402,-15085
11,10416,-3933,32149,5979,-10278,31167,-6223,-25154
12,11458,-3936,25857,7322,-10928,24227,7106,-31302
13,12500,-3933,16916,8429,-11109,14865,18191,-32765
14,13541,-3936,4588,9436,-10885,2434,27538,-29167
15,14583,-3933,-7301,9900,-10196,-9259,31672,-21194
16,15625,-3936,-18764,10028,-9243,-20214,31897,-9933
17,16666,-3933,-26342,9532,-8075,-27099,26346,2785
18,17708,-3936,-30500,8747,-6966,-30428,17611,15130
19,18750,-3933,-29070,7542,-5988,-28222,5349,25127
20,19791,-3936,-23731,6345,-5387,-22211,-6964,31339
21,20833,-3932,-13900,5098,-5157,-11985,-19009,32745
```

FIGURE 3.4   An example of data (.dat) file.

```
%Reading voltage and current input data from COMTRADE files
   %Begin
clear all;
close all;
clc;
CTR = 1;
PTR = 1;
fileID = fopen('FSLG75.cfg');
%Skip the first 2 rows of info
skip_row = textscan(fileID,'%s',3,'delimiter',',');
skip_row = textscan(fileID,'%s',3,'delimiter',',');
%The configuration info starts from the 3rd row of the.cfg
file
%Read the configuration line for IR
Data_cfg = textscan(fileID,'%f %s%s%s%s %f%f%f%f%f%f%f
%s','delimiter',',');
column_scale = Data_cfg{1,6};
IR_scale = column_scale(1);
column_set = Data_cfg{1,7};
IR_set = column_set(1);
```

```
%Read the configuration line for IA
IA_scale = column_scale(2);
IA_set = column_set(2);
%Read the configuration line for IB
IB_scale = column_scale(3);
IB_set = column_set(3);
%Read the configuration line for IC
IC_scale = column_scale(4);
IC_set = column_set(4);
%Read the configuration line for VA
VA_scale = column_scale(5);
VA_set = column_set(5);
%Read the configuration line for VB
VB_scale = column_scale(6);
VB_set = column_set(6);
%Read the configuration line for VC
VC_scale = column_scale(7);
VC_set = column_set(7);
fclose(fileID);

%Read from the .dat file
AA = importdata('FSLG75.dat');
IR = AA(:,3)*IR_scale/CTR + IR_set;
IA = AA(:,4)*IA_scale/CTR + IA_set;
IB = AA(:,5)*IB_scale/CTR + IB_set;
IC = AA(:,6)*IC_scale/CTR + IC_set;
VA = AA(:,7)*VA_scale/PTR + VA_set;
VB = AA(:,8)*VB_scale/PTR + VB_set;
VC = AA(:,9)*VC_scale/PTR + VC_set;
len = length(IA);
%End
```

3.3   CONFIGURATION OF RELAY SETTINGS

After reading input data, the next step is to configure relay settings. This process can be done by running the *Relay1_setting.m* module. The MATLAB code of this module is shown below. This module configures the impedance, directional, and current settings for a distance relay. Readers might not be familiar with some of the terminologies shown in this module, such as polarizing voltage. We will provide more detailed illustrations in later chapters.

```
%Relay1_setting.m, configuring relay settings
%Begin
Z1MAG = 18.11; %Positive-sequence line impedance magnitude
Z1ANG = 80.21*pi/180; %Positive-sequence line impedance
angle
[xx, yy] = pol2cart(Z1ANG,Z1MAG);
Z1Impedance = xx + 1i*yy;
Z0MAG = 54.3; %Zero-sequence line impedance magnitude
```

```
Z0ANG = 80.21*pi/180; %Zero-sequence line impedance angle
[xx, yy]=pol2cart(Z0ANG,Z0MAG);
Z0Impedance = xx + 1i*yy;
k0 = (Z0Impedance-Z1Impedance)/(3.0*Z1Impedance);

Inom = 5; % 5 Ampere or 1 Ampere
%Settings for directional supervision element
ZFthre = 0.49*Z1MAG;
ZRthre = 0.51*Z1MAG;
a2 = 0.1;
F50Q = 0.3;
R50Q = 0.3;

%Voltage polarization options:
%Option 1:self-polarizing, 2:cross-polarizing
Vpol_option = 1;
%Zone 1 Mho distance reach settings
Z1MG = 0.8*Z1MAG; %Mho Ground distance reach set as 80% of
Z1MAG
Z1MP = Z1MG; %Mho Phase distance reach set as 80% of Z1MAG
%Zone 1 Ground and Phase fault current minimum thresholds
Z50G1 = 5;
Z50P1 = 5;
%End
```

3.4   DATA FILTERING

The data read from the input files is called raw data. Raw data typically need to pass through data filter(s) before being used for further relay functions [3].

Cosine filter is a typical filter used for phasor-based relay functions. The principle of a cosine filter is described in Equation (3.1).

$$Ia[i] = \frac{2}{RS} \sum_{k=1}^{RS} \cos\left(\frac{2\pi}{RS}(k-1)\right) * IA[i-RS+k] \tag{3.1}$$

In Equation (3.1), IA is a vector that stores the raw data of phase A current. Ia is a vector that stores the data after filtering. The parameter RS is the rate of sampling, which equals 16 samples/cycle in this example. The index variables i and k are integers. The variable i takes value from RS to the length of the vector, and k takes value from 1 to RS. The MATLAB code of the cosine filter module is as follows:

```
%Relay1_filter.m, data filtering
%Begin
RS = 16;
filter = 1;%Enable this filter
RANGE = 4;
Inom = 5;
```

```matlab
a = -0.5 + 0.5*sqrt(3)*1i;
%{
Ir, Ia, Ib, Ic, Va, Vb, and Vc are the currents/voltages
after being processed by the filter; Then IR, IA, IB, IC,
VA, VB, VC will copy
the values from Ir, Ia, Ib, Ic, Va, Vb, Vc.
%}
if (filter==1)
    Ir = zeros(1,len);
    Ia = zeros(1,len);
    Ib = zeros(1,len);
    Ic = zeros(1,len);
    Va = zeros(1,len);
    Vb = zeros(1,len);
    Vc = zeros(1,len);

    for i = RS:len
        for k = 1:RS
            Ir(i) = Ir(i)+cos(2*pi*(k-1)/RS)*IR(i-RS+k);
        end
        Ir(i) = Ir(i)*2/RS;
        for k = 1:RS
            Ia(i) = Ia(i)+cos(2*pi*(k-1)/RS)*IA(i-RS+k);
        end
        Ia(i) = Ia(i)*2/RS;
        for k = 1:RS
            Ib(i) = Ib(i)+cos(2*pi*(k-1)/RS)*IB(i-RS+k);
        end
        Ib(i) = Ib(i)*2/RS;
        for k = 1:RS
            Ic(i) = Ic(i)+cos(2*pi*(k-1)/RS)*IC(i-RS+k);
        end
        Ic(i) = Ic(i)*2/RS;
        for k = 1:RS
            Va(i) = Va(i)+cos(2*pi*(k-1)/RS)*VA(i-RS+k);
        end
        Va(i) = Va(i)*2/RS;
        for k = 1:RS
            Vb(i) = Vb(i)+cos(2*pi*(k-1)/RS)*VB(i-RS+k);
        end
        Vb(i) = Vb(i)*2/RS;
        for k = 1:RS
            Vc(i) = Vc(i)+cos(2*pi*(k-1)/RS)*VC(i-RS+k);
        end
            Vc(i) = Vc(i)*2/RS;
    end
    IA = Ia;
    IB = Ib;
    IC = Ic;
    IR = Ir;
```

```
    VA  = Va;
    VB  = Vb;
    VC  = Vc;
end
%End
```

3.5   PHASOR CREATION AND POLARIZING VOLTAGE CALCULATION

Some protection functions are phasor-based, such as distance protection, overcurrent protection, and differential protection. Distance protection also uses polarizing voltages. In this section, we will mainly illustrate how phasors are created and how polarizing voltages are calculated in digital relays. Detailed illustration of polarizing voltages will be provided in Chapter 6.

In digital relays, the phasor of a signal (e.g., IA) is typically created using Equation (3.2). The created phasor is saved as a vector (array) of complex numbers (e.g., IAcpx).

$$IAcpx[k] = \frac{1}{\sqrt{2}}\left(IA[k] + j * IA\left[k - \frac{RS}{4}\right]\right) \quad (3.2)$$

In Equation (3.2), the k is an index variable and it takes integer values from $[RS/4]$ to the length of the vector. The parameter RS is the rate of sampling, which equals 16 samples/cycle in this example. The index $[k - RS/4]$ means the imaginary part of a phasor sample is calculated from the instantaneous value of a signal 0.25 cycles ago. The MATLAB code of this module is as follows:

```
%Relay1_phasor_vpol.m, creating phasors and calculating
Vpols
%Begin
IAcpx = zeros(1,len);
IBcpx = zeros(1,len);
ICcpx = zeros(1,len);
IRcpx = zeros(1,len);
VAcpx = zeros(1,len);
VBcpx = zeros(1,len);
VCcpx = zeros(1,len);
IABcpx = zeros(1,len);
IBCcpx = zeros(1,len);
ICAcpx = zeros(1,len);
VABcpx = zeros(1,len);
VBCcpx = zeros(1,len);
VCAcpx = zeros(1,len);
I0 = zeros(1,len);
I1 = zeros(1,len);
I2 = zeros(1,len);
V0 = zeros(1,len);
V1 = zeros(1,len);
```

```
V2 = zeros(1,len);
for v = (RS/4+1):len
    IAcpx(v) = ( IA(v)+ 1i*IA(v-RS/4) )/sqrt(2);
    IBcpx(v) = ( IB(v)+ 1i*IB(v-RS/4) )/sqrt(2);
    ICcpx(v) = ( IC(v)+ 1i*IC(v-RS/4) )/sqrt(2);
    IRcpx(v) = ( IR(v)+ 1i*IR(v-RS/4) )/sqrt(2);

    VAcpx(v) = ( VA(v)+ 1i*VA(v-RS/4) )/sqrt(2);
    VBcpx(v) = ( VB(v)+ 1i*VB(v-RS/4) )/sqrt(2);
    VCcpx(v) = ( VC(v)+ 1i*VC(v-RS/4) )/sqrt(2);

    IABcpx(v)  = IAcpx(v)-IBcpx(v);
    IBCcpx(v)  = IBcpx(v)-ICcpx(v);
    ICAcpx(v)  = ICcpx(v)-IAcpx(v);

    VABcpx(v)  = VAcpx(v)-VBcpx(v);
    VBCcpx(v)  = VBcpx(v)-VCcpx(v);
    VCAcpx(v)  = VCcpx(v)-VAcpx(v);

    I0(v) = (IAcpx(v)+IBcpx(v)+ICcpx(v))*(1.0/3);
    I1(v) = (IAcpx(v)+a*IBcpx(v)+a*a*ICcpx(v))*(1.0/3);
    I2(v) = (IAcpx(v)+a*a*IBcpx(v)+a*ICcpx(v))*(1.0/3);

    V0(v) = (VAcpx(v)+VBcpx(v)+VCcpx(v))*(1.0/3);
    V1(v) = (VAcpx(v)+a*VBcpx(v)+a*a*VCcpx(v))*(1.0/3);
    V2(v) = (VAcpx(v)+a*a*VBcpx(v)+a*VCcpx(v))*(1.0/3);
end

VpolAG = zeros(1,len);
VpolBG = zeros(1,len);
VpolCG = zeros(1,len);
VpolAB = zeros(1,len);
VpolBC = zeros(1,len);
VpolCA = zeros(1,len);
if (Vpol_option == 1) %Option 1, using self-polarizing
    VpolAG = VAcpx;
    VpolBG = VBcpx;
    VpolCG = VCcpx;
    VpolAB = VABcpx;
    VpolBC = VBCcpx;
    VpolCA = VCAcpx;

elseif (Vpol_option == 2) %Option 2, using cross-polarizing
    VpolAG = -(VBcpx + VCcpx);
    VpolBG = -(VAcpx + VCcpx);
    VpolCG = -(VAcpx + VBcpx);
    VpolAB = (-sqrt(3)*1i)*VCcpx;
    VpolBC = (-sqrt(3)*1i)*VAcpx;
    VpolCA = (-sqrt(3)*1i)*VBcpx;
end
%End
```

3.6 SUMMARY

In this chapter, we have introduced how to model digital relay functions using MATLAB. A typical MATLAB program for relay modeling consists of data reading, relay settings configuration, data filtering, protection function, and output processing modules. This chapter mainly covers the details of data reading, relay settings configuration, data filtering, and phasor creation modules. Readers are encouraged to implement the MATLAB code on their own computers. Detailed illustrations of protection function modules will be covered in later chapters.

BIBLIOGRAPHY

[1] S. E. Lyshevski, *Engineering and Scientific Computations using MATLAB*. John Wiley & Sons, 2003.
[2] C37.111-1999, IEEE Standard Common Format for Transient Data Exchange (COMTRADE) for Power Systems, DOI: 10.1109/IEEESTD.1999.90571.
[3] E. O. Schweitzer and Daqing Hou, "Filtering for protective relays," in IEEE WESCANEX93 Communications, Computers and Power in the Modern Environment-Conference Proceedings, pp. 15–23, 1993, DOI: 10.1109/WESCAN.1993.270548.

4 Overcurrent Protection

The functions for input data reading and signal processing have been illustrated in the previous chapter. Starting from this chapter, we will cover the main protection functions in digital relays. Overcurrent protection is widely used as a main protection scheme for distribution systems and an auxiliary protection scheme for transmission systems. An overcurrent protection function monitors the current flowing through a measurement point (e.g., CT). If the measured current exceeds a predetermined threshold, pertinent protection elements will be activated. Inverse-time overcurrent and instantaneous overcurrent are the two main types of overcurrent protection functions. Their principles and applications are illustrated in this chapter.

4.1 INVERSE-TIME OVERCURRENT PROTECTION

The ANSI/IEEE number for an inverse-time overcurrent protection element is 51. The characteristic of an inverse-time overcurrent function can be represented as an inverse curve, with the horizontal axis showing the measured current magnitude and the vertical axis showing the operating time [1]. The curve is to simulate the features of an electromechanical overcurrent relay whose operating time is inversely proportional to the current magnitude. The curve can be mathematically represented using Equation (4.1).

$$t = TDS \left(\frac{A}{M^P - 1} + B \right) \tag{4.1}$$

In Equation (4.1), the *TDS* (Time Dial Setting) is a parameter determined by relay engineers. The parameters A, B, and P are defined by pertinent standards enacted by the American National Standards Institute (ANSI) and the International Electrotechnical Commission (IEC) [2]. M is the ratio of the measured current magnitude (I) to the pick-up current magnitude (I_p), as shown in Equation (4.2). It should be noted that Equation (4.1) can only be used to calculate the operating time when $I \geq I_p$. When $I < I_p$, Equation (4.3) will be used to calculate a reset time.

$$M = \frac{I}{I_p} \tag{4.2}$$

$$t_{reset} = TDS \left(\frac{C}{1 - M^2} \right) \tag{4.3}$$

DOI: 10.1201/9781003629481-4

29

TABLE 4.1
Inverse-Time Overcurrent Curve Parameters

Curve Name	A	B	P	C
ANSI Moderately Inverse	0.0104	0.2256	0.02	1.08
ANSI Inverse	5.95	0.18	2.00	5.95
ANSI Very Inverse	3.88	0.0963	2.00	3.88
ANSI Extremely Inverse	5.67	0.352	2.00	5.67
IEC Class A—Standard Inverse	0.14	0.0	0.02	13.5
IEC Class B—Very Inverse	13.5	0.0	2.00	47.3
IEC Class C—Extremely Inverse	80.0	0.0	2.00	80.0

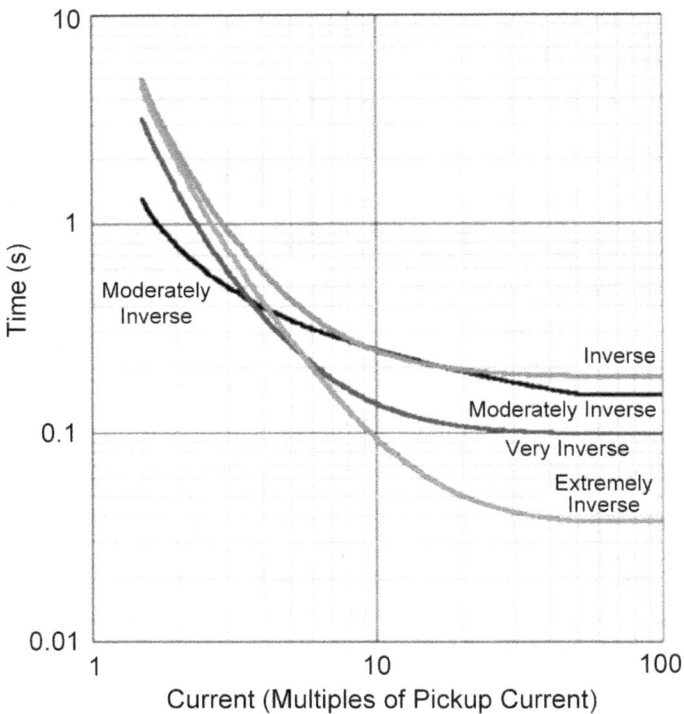

FIGURE 4.1   Comparison of inverse-time overcurrent curves when TDS = 1.0.

Parameters of A, B, P, and C according to ANSI and IEC standards are listed in Table 4.1.

A comparison of ANSI inverse, moderately inverse, very inverse, and extremely inverse curves when $TDS = 1.0$ is shown in Figure 4.1.

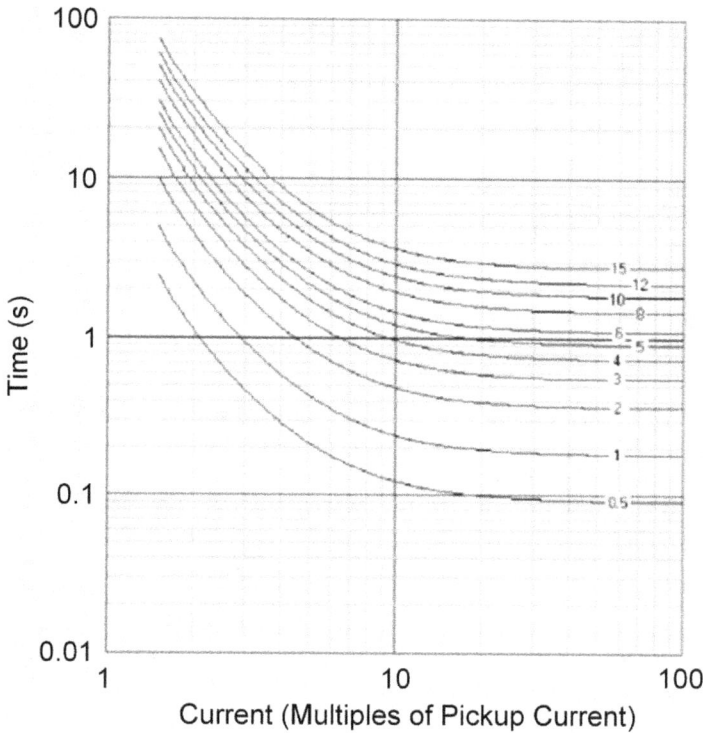

FIGURE 4.2 A group of curves with the same shape and different TDS values.

The coordination between upstream and downstream overcurrent relays can be achieved by using the same shape of curves with different TDS values [3]. Figure 4.2 shows a group of ANSI inverse curves with different TDS values. The desired coordination can be accomplished by increasing the time dial settings as one proceeds toward the source. Fault analysis is usually performed before configuring relay settings. The pickup current must be set at or below the minimum fault current and above the maximum load current, usually with a margin around both. An example of overcurrent relay coordination is illustrated in Example 4.1.

Example 4.1

The one-line diagram of a 3-bus distribution system is shown in Figure 4.3. The source voltage $V_S = 1.0$ per unit (pu). The positive- and zero-sequence source impedances are $Z_{S1} = j0.5$ pu and $Z_{S0} = j0.2$ pu. For the feeder section from Bus 0 to Bus 1, the positive- and zero-sequence impedances are $Z_{fd11} = j1.0$ pu and $Z_{fd10} = j1.5$ pu. For the feeder section from Bus 1 to Bus 2, the positive and zero-sequence impedances are $Z_{fd21} = j3.0$ pu and $Z_{fd20} = j4.5$ pu. For each section, the negative-sequence impedance equals its positive-sequence impedance. The base

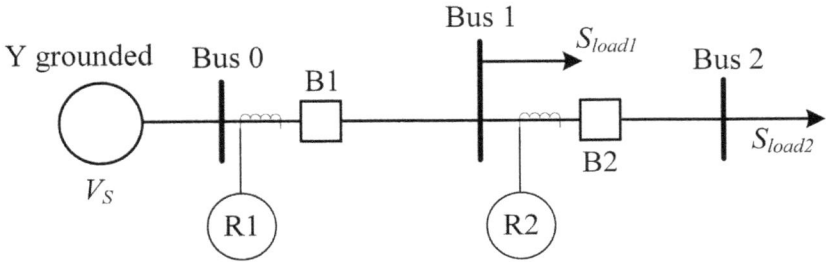

FIGURE 4.3 One-line diagram for Example 4.1.

values are $V_{llbase} = 24$ kV and S_{base} (3-phase) = 100 MVA. $S_{load1} = 3.5$ MVA at unity power factor. $S_{load2} = 3$ MVA at unity power factor. The desired coordinating time interval is set as 18 cycles, which equals 0.3 second. Determine the pickup currents and time dial settings for the two overcurrent relays R1 and R2.

Solution:

The first step is to perform a fault current calculation. This can be achieved by using a fault simulation program or using analytical methods. Pertinent theories are illustrated in detail in Appendix A1 of this book.

After performing fault current calculation, we find the load currents, $I_{ld1} = 84.2$ A and $I_{ld2} = 72.17$ A.

The fault currents for a fault occurring at Bus 1 are calculated as below:
$I_{3ph\_bus1} = 1603.75$ A, $I_{LL\_bus1} = 1388.89$ A, $I_{SLG\_bus1} = 1535.51$ A.

The fault currents for a fault occurring at Bus 2 are calculated as below:
$I_{3ph\_bus2} = 534.58$ A, $I_{LL\_bus2} = 462.96$ A, $I_{SLG\_bus2} = 474.79$ A.

We choose to use a very inverse characteristic; the equation for calculating the pickup time t_{pu} is shown below:

$$t_{pu} = TDS\left(\frac{3.88}{M^2 - 1} + 0.0963\right)$$

Determine the CT ratio for each relay location:
- Total load current at Relay 1 (controlling circuit breaker B1):

  $I_{ldR1} = I_{ld1} + I_{ld2} = 156.37$ A.

  Therefore, we can set the CT ratio, $CTR_{R1} = 160/5$.
- Total load current at Relay 2 (controlling circuit breaker B2):

  $I_{ldR2} = I_{ld2} = 72.17$ A.

  Therefore, we can set the CT ratio, $CTR_{R2} = 75/5$.

The next step is to determine the settings for Relay 2.
- First, determine the load current on CT secondary.

$I_{ldR2\_sec} = I_{ldR2} / CTR_{R2} = 4.81 \, \text{A}.$

The minimum phase fault current seen by Relay 2:

$I_{ph\_f\_minR2\_sec} = I_{LL\_bus2} / CTR_{R2} = 30.86 \, \text{A}.$

We can set the pickup current to be approximately between twice maximum load current and half of minimum fault current.

$$2I_{ldR2\_sec} = 9.62 \, \text{A and} \, 0.5I_{ph\_f\_minR2\_sec} = 15.43 \, \text{A}.$$

We can set the Relay 2 pickup current $I_{pu\_R2} = 10$ A, and the time dial setting for Relay 2 at the minimum $TDS_{R2} = 0.5$.

It should be noted that in this case, $0.5I_{ph\_f\_minR2\_sec}$ is greater than $2I_{ldR2\_sec}$; therefore, the selection of pickup current is straightforward. If $0.5I_{ph\_f\_minR2\_sec}$ is smaller than $2I_{ldR2\_sec}$, some adjustments need to be made depending on specific circumstances.

Now we calculate the Relay 2 pickup time for the maximum fault current, which corresponds to a three-phase fault occurring right at the Relay 2 location. This fault current value approximately equals a three-phase fault occurring at Bus 1 because Relay 2 is located very close to Bus 1.

$$I_{ph\_f\_maxR2\_sec} = I_{3ph\_bus1} / CTR_{R2} = 106.92 \, \text{A}.$$

The corresponding M value is $M = I_{ph\_f\_maxR2\_sec} / I_{pu\_R2} = 10.69$.

By plugging in the M and TDS_{R2} values to the very inverse characteristic equation, we can find the Relay 2 pickup time for a maximum fault current.

$$t_{pu\_R2\_lf\_max} = 0.065 \, \text{s}.$$

Now we need to set Relay 1 (controlling B1).
- First, determine the load current on CT secondary.

$I_{ldR1\_sec} = I_{ldR1} / CTR_{R1} = 4.89 \, \text{A}.$

- Minimum phase fault current seen by Relay 1, which still corresponds to a fault at Bus 2.

$I_{ph\_f\_minR1\_sec} = I_{LL\_bus2} / CTR_{R1} = 14.47 \, \text{A}.$

We can set the pickup current to be approximately between twice the maximum load current and half of the minimum fault current.

$$2I_{ldR1\_sec} = 9.78 \, \text{A and} \, 0.5I_{ph\_f\_minR1\_sec} = 7.23 \, \text{A}.$$

In this case, $2I_{ldR1\_sec}$ is greater than $0.5I_{ph\_f\_minR1\_sec}$ and we need to adjust the coefficient 2 to 1.5.

$1.5I_{ldR1\_sec} = 7.33$ A, which is still greater than $0.5I_{ph\_f\_minR1\_sec}$.

We may set the Relay 1 pickup current $I_{pu\_R1} = 7.33$ A. It should be noted that in field applications, this coefficient could be further adjusted.

Now, we need to choose a time dial setting based on the coordinating time interval. We already calculated Relay 2 pickup time for a maximum fault current $t_{pu\_R2\_lf\_max} = 0.065\,\mathrm{s}$.

For the same fault current, the desired minimum pickup time for Relay 1 would be:

$$t_{pu\_R1\_desired} = t_{pu\_R2\_lf\_max} + \text{coordinating time interval } (\text{CTI}) = 0.365\,\mathrm{s}.$$

The M value for Relay 1 for this fault is $M = \left(I_{3ph\_bus1} / CTR_{R1}\right) / I_{pu\_R1} = 6.84$.

Plugging in the pickup time $t_{pu\_R1\_desired}$ and M values to the very inverse characteristic equation, we can calculate the time dial setting (TDS) for Relay 1.

The calculated TDS for Relay 1 is 2.02. Set it to the nearest 0.1, $TDS_{R1} = 2.1$.

Next, we check the coordination when the fault current is minimum, which corresponds to a line-to-line (LL) fault occurring at Bus 2.

For Relay 1, the M value is $\left(I_{LL\_bus2} / CTR_{R1}\right) / I_{pu\_R1} = 1.97$.

For Relay 2, the M value is $\left(I_{LL\_bus2} / CTR_{R2}\right) / I_{pu\_R2} = 3.09$.

By plugging in the M and TDS values to the very inverse characteristic equation for Relay 1 and Relay 2, respectively, we can find that their pickup time difference is 2.74 s, which meets the criteria.

4.2   INSTANTANEOUS OVERCURRENT PROTECTION

When a fault current magnitude is high enough, it is more preferable for an overcurrent relay to respond instantaneously to the fault instead of using an inverse curve to determine a waiting time. Adding an instantaneous overcurrent element could accelerate the tripping process for a high-current fault.

Let us look at Example 4.1 again. Now, we add an instantaneous overcurrent protection function to Relay 1. In this example, we consider the current magnitude corresponding to a three-phase fault occurring at 60% of the feeder length to be high enough to trigger the instantaneous overcurrent function. After performing fault analysis, this triggering current is calculated as $I_{f3ph60\%} = 2186.93$ A.

Converting this triggering current to a secondary value: $I_{f3ph60\%\_sec} = I_{f3ph60\%} / CTR_{R1} = 68.34\,\mathrm{A}$

Round this value up and use this as the pickup setting for the 50P (instantaneous phase) element.

$$I_{R1\_50\_set} = 70 \text{ A}.$$

This instantaneous overcurrent element can be combined with the inverse-time overcurrent element. The pickup time for Relay 1 can be described by the following equation:

$$t_{pu\_R1} = \begin{cases} using\ the\ inverse - time\ overcurrent\ equation,\ if\ 0 \leq M \leq \dfrac{I_{R1\_50\_set}}{I_{pu\_R1}} \\ \\ 0.0001s,\ if\ M > \dfrac{I_{R1\_50\_set}}{I_{pu\_R1}} \end{cases}$$

The overcurrent characteristic curves for Relay 1 and Relay 2 are shown in Figure 4.4.

An overcurrent protection element typically calculates current phasors (using Equation 3.2) first, then compares the phasor values to the characteristic curves. It should be noted that in industry applications, an instantaneous overcurrent element will not immediately pick up if one phasor current sample is above $I_{Relay\_50\_set}$. Before a pickup decision is confirmed, typically 3–5 continuous phasor current

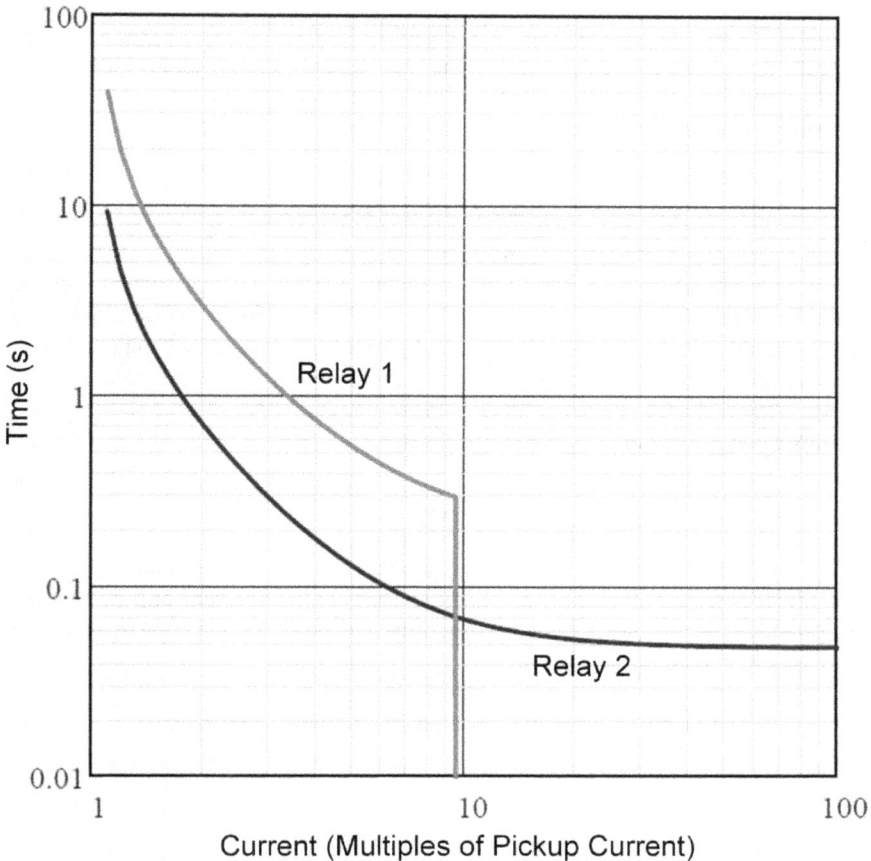

FIGURE 4.4 Overcurrent characteristic curves for the two relays.

DC Bus +

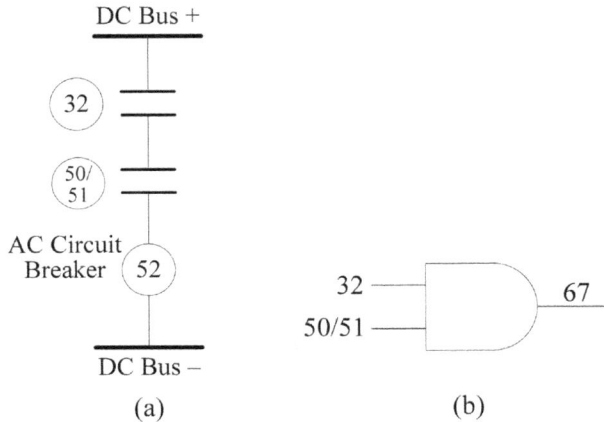

(a)   (b)

FIGURE 4.5 Directional overcurrent diagram for (a) electromechanical relays and (b) digital relays.

samples must stay above $I_{Relay\_50\_set}$. This mechanism helps to reduce the likelihood of misoperation caused by transients, and therefore improves the overall security of protection functions.

4.3 DIRECTIONAL OVERCURRENT PROTECTION

The ANSI/IEEE number for a directional overcurrent protection element is 67. It is the combination of a directional element (32) and an overcurrent element (50 or 51). The diagram for an electromechanical relay with directional and overcurrent coils is shown in Figure 4.5 (a). In digital relays, functions can be achieved by programming. The logic diagram for a directional overcurrent function in a digital relay is shown in Figure 4.5 (b). The principles of directional elements will be illustrated in the next chapter of this book.

4.4 MATLAB IMPLEMENTATION

In this section, we will illustrate the implementation of an instantaneous overcurrent protection function using MATLAB. The system diagram and parameters are provided in Example 4.1. The functions are developed for the overcurrent relay R1 located at Bus 0.

The *Relay_main.m* is the master file. Running this file will run the entire program. The MATLAB code of the *Relay_main.m* file is as follows:

```
%Relay_main.m, the master file of the program
%Begin
clear all;
close all;
clc;
%Read Phase ABC current data at Bus 0
```

```
Relay_readdata;
%Configure the Relay settings
Relay_setting;
%Process the input currents through a filter
Relay_filter;
%Create phasors
Relay_phasor;
%The calculation and trip logics for instantaneous
overcurrent protection
Relay_50_Bus0;
%Plot the results
Relay_plot;
%End
```

The *Relay_readdata.m* reads the input currents. In this example, the input data are saved in a MATLAB data file *data_ch4.mat*. The fault event corresponds to a phase-A-to-ground fault occurring at $t = 0.055$ second. The location of the fault is 30% distance from Bus 0 to Bus 1. The MATLAB code of the *Relay_readdata.m* file is shown below. The data are resampled to 16 samples per cycle. The CT ratio (CTR) is 160/5, as illustrated in Example 4.1.

```
%Relay_readdata.m, reading the input currents
%Begin
CTR = 160.0/5;
VIread = load('data_ch4.mat');
Ia = VIread.iBus0faBus1fa/CTR;%Phase A current
Ib = VIread.iBus0fbBus1fb/CTR;%Phase B current
Ic = VIread.iBus0fcBus1fc/CTR;%Phase C current
time_before = VIread.t;
dt_before = time_before(2)-time_before(1);
RS_before = round( (1*3.0)/(60*dt_before) );
RS_after = 16*3;
%Resample the data to 16 samples per cycle
IA = resample(Ia, RS_after, RS_before);
IB = resample(Ib, RS_after, RS_before);
IC = resample(Ic, RS_after, RS_before);
len = length(IA);
%End
```

The *Relay_setting.m* configures the relay settings. The MATLAB code of the *Relay_setting.m* file is shown below. As illustrated in Example 4.1, the instantaneous overcurrent (50) setting for the relay R1 at Bus 0 is 70 A (secondary).

```
%Relay_setting.m, relay settings
%Begin
Ithreshold = 70;%Current threshold
for i = 1:len
    Ithre(i) = Ithreshold;
end
%End
```

The *Relay_filter.m* performs a filter algorithm to the currents. The MATLAB code is as follows:

```
%Relay_filter.m
%Begin
RS = 16;
filter = 1;
%Filter, if (filter==0), Filter will be skipped
%{
IA, IB, IC are the
currents after being processed by the filter
%}
if (filter==1)
    Ia = zeros(len,1);
    Ib = zeros(len,1);
    Ic = zeros(len,1);

    for i = RS:len
        for k = 1:RS
            Ia(i) = Ia(i)+cos(2*pi*(k-1)/RS)*IA(i-RS+k);
        end
        Ia(i) = Ia(i)*2/RS;

        for k = 1:RS
            Ib(i) = Ib(i)+cos(2*pi*(k-1)/RS)*IB(i-RS+k);
        end
        Ib(i) = Ib(i)*2/RS;

        for k = 1:RS
            Ic(i) = Ic(i)+cos(2*pi*(k-1)/RS)*IC(i-RS+k);
        end
        Ic(i) = Ic(i)*2/RS;
    end
    IA = Ia;
    IB = Ib;
     IC = Ic;
end
%End
```

The *Relay_phasor.m* computes phasors for further protection functions. The MATLAB code is as follows:

```
%Relay_phasor.m, phasor creation
%Begin
IAcpx = zeros(1,len);
IBcpx = zeros(1,len);
ICcpx = zeros(1,len);

for v = (RS/4+1):len
    IAcpx(v) = ( IA(v)+ 1i*IA(v-RS/4) )/sqrt(2);
```

```
    IBcpx(v) = ( IB(v)+ 1i*IB(v-RS/4) )/sqrt(2);
    ICcpx(v) = ( IC(v)+ 1i*IC(v-RS/4) )/sqrt(2);
end
%End
```

The *Relay_50_Bus0.m* is the instantaneous overcurrent function for the relay R1 at Bus 0. The MATLAB code is shown below. In this example, the pick-up logic value is programmed to change to 1 instantaneously when a sample of current magnitude exceeds the threshold. In commercial relays, the pick-up logic value is typically programmed to change to 1 when at least 3 to 5 continuous samples exceed the threshold. This mechanism helps reduce the possibility of misoperation caused by noise.

```
%Relay_50_Bus0.m
%Calculation and trip logics for the instantaneous
overcurrent protection
%For the relay R1 at Bus 0;
%Begin
%The pick-up and trip logics
OC50A_pu = zeros(1,len);%Phase A pick-up logic;
OC50B_pu = zeros(1,len);%Phase B pick-up logic;
OC50C_pu = zeros(1,len);%Phase C pick-up logic;
OC50_trip = zeros(1,len);%Relay 1 instantaneous overcurrent
trip logic;
for i = 1:len
    if (abs(IAcpx(i))>Ithreshold)
        OC50A_pu(i) = 1;
    end
    if (abs(IBcpx(i))>Ithreshold)
        OC50B_pu(i) = 1;
    end
    if (abs(ICcpx(i))>Ithreshold)
        OC50C_pu(i) = 1;
    end
    OC50_trip(i) = OC50A_pu(i)||OC50B_pu(i)||OC50C_pu(i);
end
%End
```

The *Relay_plot.m* plots the results. The MATLAB code is as follows:

```
%Relay_plot.m, plotting the results
%Begin
t = zeros(len,1);
freq = 60;
for i = 1:len
    t(i) = (i-1)*1000.0/(freq*RS);
end

figure;
```

```
plot(t,abs(IAcpx),'b',t,abs(IBcpx),'r',t,abs(ICcpx),'k',t,It
hre,'-- m');
xlabel('Time (ms)');
ylabel('Current (A)');

figure;
plot(t,OC50A_pu,'b',t,OC50B_pu,'r',t,OC50C_pu,'k');
xlabel('Time (ms)');
ylabel('Logic Value');
ylim([-0.1 1.1]);

figure;
plot(t,OC50_trip,'k');
xlabel('Time (ms)');
ylabel('Logic Value');
ylim([-0.1 1.1]);
%End
```

The secondary side three-phase current magnitudes are shown in Figure 4.6 (a). The three-phase instantaneous overcurrent (50) pickup logic values are shown in Figure 4.6 (b).

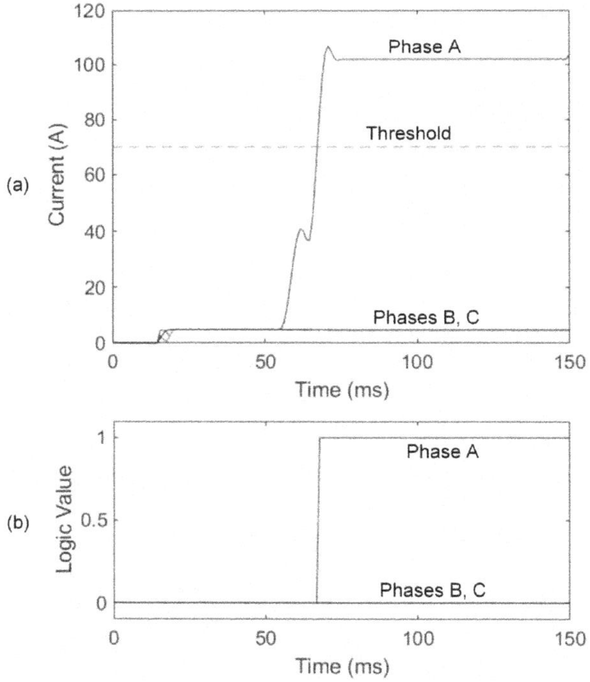

FIGURE 4.6   (a) Three-phase current magnitudes and (b) three-phase pick-up logic values.

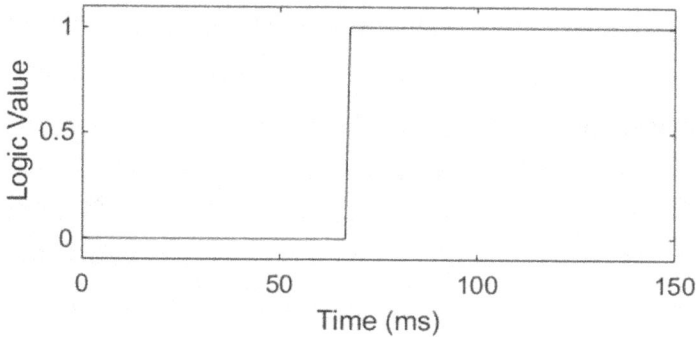

FIGURE 4.7   Overall trip logic value of the instantaneous overcurrent function.

The overall trip logic value, which is the combination of the three-phase pick-up logic values, is shown in Figure 4.7.

4.5   SUMMARY

In this chapter, the principles of overcurrent protection functions have been illustrated. Overcurrent protection mainly includes inverse-time and instantaneous types. Combined with a directional element, an overcurrent element becomes a directional overcurrent element. The MATLAB implementation of an instantaneous overcurrent protection function has been illustrated in this chapter. Readers are encouraged to develop the MATLAB code for inverse-time overcurrent and directional overcurrent functions. Principles of directional elements will be illustrated in the next chapter.

4.6   PROBLEMS

Problem 4.1

For the distribution system described in Example 4.1. The positive- and zero-sequence source impedances are $Z_{s1} = j0.3$ pu and $Z_{s0} = j0.1$ pu. For the feeder section from Bus 0 to Bus 1, the positive- and zero-sequence impedances are $Z_{fd11} = j1.5$ pu and $Z_{fd10} = j2.0$ pu. For the feeder section from Bus 1 to Bus 2, the positive- and zero-sequence impedances are $Z_{fd21} = j2.0$ pu and $Z_{fd20} = j3.0$ pu. Other parameters are the same as Example 4.1. Analytically calculate the fault current values for single-line-to-ground (SLG), line-to-line (LL), and three-phase (3PH) faults occurring at Bus 1 and Bus 2, respectively.

Problem 4.2

Based on the results calculated in Problem 4.1, use an ANSI moderately inverse curve with a desired coordinating time interval (CTI) of 20 cycles, and determine the CT ratio, pickup current, and time dial settings for Relay 1 and Relay 2 with reasonable judgment.

BIBLIOGRAPHY

[1] J. C. Das, *Power System Protective Relaying*. CRC Press, 2017.

[2] P. M. Anderson, C. Henville, R. Rifaat, B. Johnson, and S. Meliopoulos, *Power System Protection*, 2nd Ed. Wiley, 2022.

[3] R. W. Wall and B. K. Johnson, "Using TACS functions within EMTP to teach protective relaying fundamentals," *IEEE Transactions on Power Systems*, vol. 12, no. 1, pp. 3–10, 1997.

5 Directional Elements

The concept of directional elements has been briefly introduced in the previous chapter. The purpose of using a directional element is to check whether or not a fault is in the forward direction. A main protection element, such as an overcurrent or a distance element, should only pick up when a forward direction is confirmed. This mechanism is called directional supervision, and it provides additional security to protection systems. The principle of determining a fault direction is typically based on comparing the phase angle difference between a measured voltage and current. In this chapter, we illustrate several methods that are commonly used in digital relays for fault direction determination.

5.1 MAXIMUM TORQUE ANGLE

The principle of determining a fault direction is typically based on comparing the phase angle difference between a measured voltage and current [1]. This can be illustrated with the power system shown in Figure 5.1.

The power system shown in Figure 5.1 consists of two transmission lines, two voltage sources, and two loads. The voltage sources represent the equivalent sources of two adjacent systems. The relay R1 constantly takes voltage and current measurements. V_{R1} and I_{R1} are the phase A voltage and current measured by R1. We use voltage V_{R1} as the phase angle reference. The phasor diagram is shown in Figure 5.2. In normal conditions, the phase A load current is in the direction of either $I_{R1\_load\_fwd}$ or $I_{R1\_load\_rvs}$, depending on the load flow direction.

Let us use θ_{ZL11} to represent the positive-sequence impedance angle of transmission line 1 (i.e., the transmission line between R1 and R2), and use θ_{ZL21} to represent the positive-sequence impedance angle of transmission line 2 (i.e., the transmission line between V_{R1} and V_{S1}).

If a metallic fault involving phase A occurs at Location 1, the fault is in the forward direction from R1's perspective. The measured fault current $I_{R1\_fault\_fwd}$ will lag the voltage V_{R1} by θ_{ZL11}, which is approximately 75–85 degrees. If a metallic fault involving phase A occurs at Location 2, the fault is in the reverse direction from R1's perspective. The measured fault current $I_{R1\_fault\_rvs}$ will lead the voltage V_{R1} by $(180°- \theta_{ZL21})$, which is approximately 95–105 degrees.

In a transmission line relay, there is a parameter θ_{MTA} called the maximum torque angle. A relay typically uses the positive-sequence impedance angle of the protected transmission line as the maximum torque angle. In the example shown in Figure 5.1, the relay R1 can use θ_{ZL11} as the maximum torque angle. This angle is a key parameter for directional elements.

DOI: 10.1201/9781003629481-5

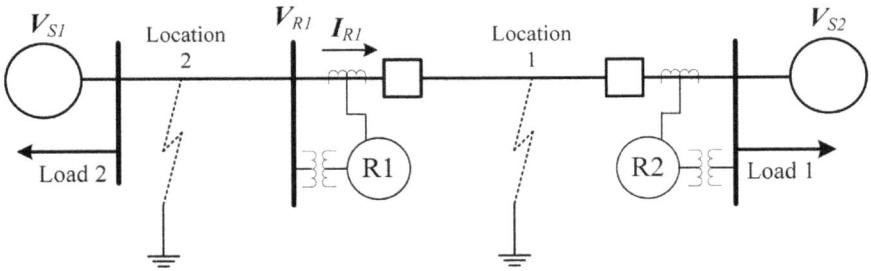

FIGURE 5.1 A power system for illustrating fault direction determination.

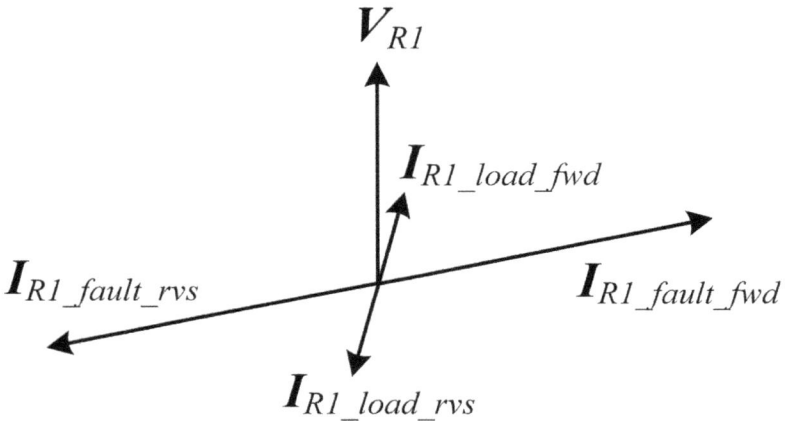

FIGURE 5.2 Phasor diagram for relay voltage and current.

5.2 TORQUE-BASED DIRECTION DETERMINATION

Torque-based direction determination methods compute the product of a phase voltage and a phase current with correction of the maximum torque angle using Equation (5.1). The computed value is called "torque", which is a term coming from electromechanical relays.

$$T_A = |I_A| \cdot |V_{AN}| \cdot \cos\left(\theta_{VAN} - \theta_{IA} - \theta_{MTA}\right) \tag{5.1}$$

In Equation (5.1), the subscripts "A" and "AN" refer to phase A. T_A is the computed torque value. I_A is phase A current and V_{AN} is phase-A-to-ground voltage. θ_{VAN} is the phase angle of V_{AN} and θ_{IA} is the phase angle of I_A. θ_{MTA} is the maximum torque angle. The computed torque is then compared to a predefined torque threshold T_{thd} to determine the fault direction. If the computed torque is greater than the threshold, then it confirms a forward direction in phase A. This equation can also be applied for phases B and C.

Bus S $\quad V_{R1} \quad I_{R1} \quad$ Bus R

V_{S1}

R1

V_{S2}

Line 1 $\qquad$ Line 2

Bus 1

345 kV   345 kV
100 MVA   100 MVA

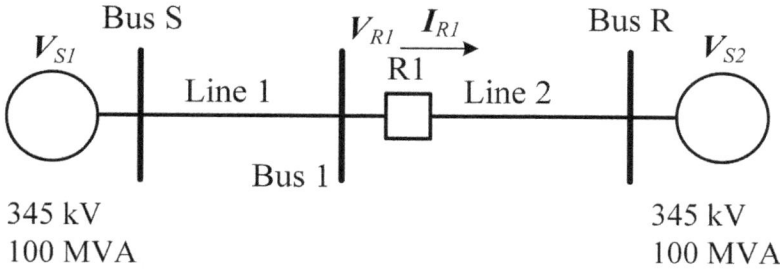

FIGURE 5.3 The diagram for Example 5.1.

Example 5.1

The one-line diagram of a power system is shown in Figure 5.3. The relay located at R1 has a directional element with the forward direction set to look into Line 2. The CT ratio for the relay is 1500/5, and the VT ratio for the relay is 345 kV/120 V. The positive-sequence impedance of Line 2 is 94.22 Ω primary (or 9.94 Ω secondary), with 85 degrees. In a fault event, the three-phase currents and voltages seen by the relay (secondary side) are listed below.

$I_{R1\_A} = 2.937\ A,\ with - 85.79\ degree$ $\qquad$ $V_{R1\_AN} = 21.89\ V,\ with - 0.79\ degree$
$I_{R1\_B} = 2.937\ A,\ with - 205.79\ degree$ $\qquad$ $V_{R1\_BN} = 21.89\ V,\ with - 120.79\ degree$
$I_{R1\_C} = 2.937\ A,\ with + 34.21\ degree$ $\qquad$ $V_{R1\_CN} = 21.89\ V,\ with + 119.21\ degree$

Use the torque method to determine the direction of the fault (forward or reverse). You may use 25 VA (secondary) as the threshold T_{thd}.

Solution:

We can set the maximum torque angle $\theta_{MTA} = 85°$.

The calculated phase A torque:
$$T_A = \left|I_{R1\_A}\right| \cdot \left|V_{R1\_AN}\right| \cdot \cos\left(\theta_{VAN} - \theta_{IA} - \theta_{MTA}\right) = 64.29\ \text{VA} > T_{thd.}$$

The calculated phase B torque:
$$T_B = \left|I_{R1\_B}\right| \cdot \left|V_{R1\_BN}\right| \cdot \cos\left(\theta_{VBN} - \theta_{IB} - \theta_{MTA}\right) = 64.29\ \text{VA} > T_{thd.}$$

The calculated phase C torque:
$$T_C = \left|I_{R1\_C}\right| \cdot \left|V_{R1\_CN}\right| \cdot \cos\left(\theta_{VCN} - \theta_{IC} - \theta_{MTA}\right) = 64.29\ \text{VA} > T_{thd.}$$

All three calculated torques are positive and larger than the threshold.

Using the three-phase currents, we obtain the positive-, negative-, and zero-sequence current magnitudes are 2.937 A, 0 A, and 0 A, respectively. These results indicate that it is a three-phase fault in the forward direction.

5.3 SEQUENCE IMPEDANCE-BASED DIRECTION DETERMINATION

Sequence impedance-based methods determine the direction of a fault by calculating the negative-, zero-, or positive-sequence impedance [2]. The negative-sequence impedance-based method is typically preferred in applications because negative-sequence current exists in most types of faults, and it is not affected significantly by load current. If there is an insufficient amount of negative-sequence current, then a zero-sequence impedance-based method is typically preferred over a positive-sequence impedance-based method because a positive-sequence current is often affected by load current.

The negative-sequence impedance Z_2 can be calculated using Equation (5.2).

$$Z_2 = \frac{Re\left(V_2 \cdot \overline{I_2 \cdot 1\angle\theta_{MTA}}\right)}{\left|I_2\right|^2} \tag{5.2}$$

where V_2 is the negative-sequence voltage, I_2 is the negative-sequence current, and θ_{MTA} is the maximum torque angle. The bar means taking the conjugate of a complex number. The "Re" means taking the real part of a complex number. The calculated Z_2 is compared with a forward direction threshold $ZF2_{thd}$ and a reverse direction threshold $ZR2_{thd}$ to determine the direction of the fault. If the calculated Z_2 is smaller than $ZF2_{thd}$, it indicates a forward direction. If the calculated Z_2 is greater than $ZR2_{thd}$, it indicates a reverse direction.

It should be noted that before using the negative-sequence-based directional element, a qualification test must be performed to ensure that there is a sufficient amount of negative-sequence current. If Equation (5.3) is true, then the negative-sequence-based directional element can be used. The I_1 is the positive-sequence current. The parameter a_2 is a positive real number (e.g., 0.1 or 0.2) set by relay engineers.

$$\frac{\left|I_2\right|}{\left|I_1\right|} > a_2 \tag{5.3}$$

Similarly, the zero-sequence impedance Z_0 can be calculated using Equation (5.4).

$$Z_0 = \frac{Re\left(V_0 \cdot \overline{I_0 \cdot 1\angle\theta_{MTA}}\right)}{\left|I_0\right|^2} \tag{5.4}$$

Where V_0 is the zero-sequence voltage, I_0 is the zero-sequence current, and θ_{MTA} is the maximum torque angle. The calculated Z_0 is compared with a forward direction threshold $ZF0_{thd}$ and a reverse direction threshold $ZR0_{thd}$ to determine the direction of the fault. If the calculated Z_0 is smaller than $ZF0_{thd}$, it indicates a forward direction. If the calculated Z_0 is greater than $ZR0_{thd}$, it indicates a reverse direction.

It should be noted that before using the zero-sequence-based directional element, a qualification test must be performed to ensure that there is a sufficient amount of zero-sequence current. If Equation (5.5) is true, then the zero-sequence-based directional element can be used. The parameter a_0 is a positive real number (e.g., 0.1 or 0.2) set by relay engineers.

$$\frac{|I_0|}{|I_1|} > a_0 \tag{5.5}$$

If a fault event does not provide a sufficient amount of negative- and zero-sequence currents, we may use a positive-sequence impedance to determine the fault direction. The positive-sequence impedance Z_1 is calculated using Equation (5.6).

$$Z_1 = \frac{V_1}{I_1} \tag{5.6}$$

Please note that the result of Equation (5.6) is a complex number, whose angle is θ_{Z1}. If θ_{Z1} satisfies Equation (5.7), then it indicates a forward direction. Otherwise, it indicates a reverse direction.

$$-90^\circ + \theta_{MTA} < \theta_{Z1} < 90^\circ + \theta_{MTA} \tag{5.7}$$

Example 5.2

For the power system shown in Figure 5.3, the three-phase currents and voltages seen by the relay (secondary side) in a fault event are listed as follows.

$I_{R1\_A} = 1.838\ A,\ with - 76.96\ degree$ $\qquad V_{R1\_AN} = 21.89\ V,\ with - 0.79\ degree$

$I_{R1\_B} = 0.718\ A,\ with - 126.67\ degree$ $\qquad V_{R1\_BN} = 78.09\ V,\ with - 142.03\ degree$

$I_{R1\_C} = 0.718\ A,\ with + 113.33\ degree$ $\qquad V_{R1\_CN} = 80.09\ V,\ with + 121.48\ degree$

Use negative- and zero-sequence impedance-based methods to determine the direction of the fault (forward or reverse). When using the negative-sequence impedance-based method, you may use $a_2 = 0.2$, $ZF2_{thd} = 4.9\ \Omega$, and $ZR2_{thd} = 5.1\ \Omega$. When using the zero-sequence impedance-based method, you may use $a_0 = 0.2$, $ZF0_{thd} = 4.9\ \Omega$, and $ZR0_{thd} = 5.1\ \Omega$.

Solution:

From the three-phase currents, we can obtain the negative-, zero-, and positive-sequence currents and voltages as follows:

$I_2 = 0.578\ A,\ with - 99.92\ degree$ $\qquad V_2 = 9.375\ V,\ with + 165.04\ degree$

$I_0 = 0.578\ A,\ with - 99.92\ degree$ $\qquad V_0 = 28.021\ V,\ with + 165.84\ degree$

$I_1 = 0.896\ A,\ with - 46.76\ degree$ $\qquad V_1 = 58.799\ V,\ with - 8.78\ degree$

Based on the symmetrical components, this is a phase-A-to-ground (AG) fault without fault resistance but with significant load flow. Because I_2 and I_0 are in phase and identical in magnitude, $|I_1|$ is large than $|I_2|$ and $|I_0|$.

$\dfrac{|I_2|}{|I_1|} = 0.64 > a_2$, which indicates that there is sufficient negative-sequence current. Thus, we can use the negative-sequence impedance-based method.

The negative-sequence impedance $Z_2 = \dfrac{Re\left(V_2 \cdot \overline{I_2 \cdot 1 \angle \theta_{MTA}}\right)}{|I_2|^2} = -16.22\ \Omega < ZF2_{thd}$. It indicates a forward fault.

$\dfrac{|I_0|}{|I_1|} = 0.64 > a_0$, which indicates that there is sufficient zero-sequence current. Thus, we can use the zero-sequence impedance-based method.

The zero-sequence impedance $Z_0 = \dfrac{Re\left(V_0 \cdot \overline{I_0 \cdot 1 \angle \theta_{MTA}}\right)}{|I_0|^2} = -48.49\Omega < ZF0_{thd}$ indicates a forward fault.

It should be noted that in this example, $ZF2_{thd}$, $ZR2_{thd}$, $ZF0_{thd}$, and $ZR0_{thd}$ were given as constants. In field applications, these parameters could be given as variables whose values are calculated dynamically from certain equations specified by relay vendors.

5.4   SUMMARY

In this chapter, several commonly used algorithms for fault direction determination have been illustrated. Directional elements are a type of supervising elements that provide additional security and selectivity to the main protection functions.

5.5   PROBLEMS

Problem 5.1

For the power system and fault event described in Example 5.1, use the positive-sequence impedance-based method to determine the fault direction. Can we use negative- and zero-sequence impedance-based methods to determine the fault direction? Why?

Problem 5.2

For the power system and fault event described in Example 5.2, use the torque-based method to determine the fault direction. What challenges are encountered in the determination process? In applications, sequence impedance-based methods, especially negative- and zero-sequence-based methods, are typically preferred over torque-based methods.

BIBLIOGRAPHY

[1] P. M. Anderson, C. Henville, R. Rifaat, B. Johnson, and S. Meliopoulos, *Power System Protection*, 2nd Ed. Wiley, 2022.

[2] R. Lavorin, D. Hou, H. J. Altuve, N. Fischer, and F. Calero, "Selecting directional elements for impedance-grounded distribution systems," in the 34th Annual Western Protective Relay Conference, Spokane, WA, 2007. https://selinc.com/api/download/3499/

6 Distance Protection

Distance protection, also known as impedance protection, is widely used for the protection of transmission lines. A distance element calculates an effective impedance using voltage and current measurements, and compares this calculated impedance with a predefined area on an impedance plane to determine whether or not to pick up. Because of the utilization of both voltage and current measurements, a distance element generally provides better sensitivity and selectivity compared to an overcurrent element. Depending on applications, the predefined area can be a circle or a quadrilateral area. In this chapter, we illustrate the principle of distance protection and the implementation of distance protection algorithms using MATLAB.

6.1 DISTANCE PROTECTION INTRODUCTION

The ANSI number for distance protection is 21. In a transmission line protective relay, a distance element typically consists of ground-distance (21G) and phase-distance (21P) elements [1]. A ground-distance element includes phase-A-to-ground (AG), phase-B-to-ground (BG), and phase-C-to-ground (CG) elements. A phase-distance element includes phase-A-to-B (AB), phase-B-to-C (BC), and phase-C-to-A (CA) elements. The logic diagram of a Zone 1 distance element is shown in Figure 6.1. It should be noted that this is a simplified logic diagram. In commercial relays, there are more supervising elements, such as single-pole-open checking [2], which are not the main focus of this chapter and therefore not shown in this diagram.

In Figure 6.1, the Zone 1 ground-distance (Z1G) element picks up when at least one of the Zone 1 Mho ground-distance elements (e.g., MAG1F) picks up. The "MAG1F" means Zone 1 mho ground-distance element for AG faults, in which the letter "F" represents forward direction. This element picks up only if all the conditions on the left side of the AND gate are true. The "mAG1 < Z1MG" is the main condition, which represents that the impedance calculated by the relay is within the predefined Zone 1 setting. This main condition is the core principle for a distance element. All other conditions below the main condition are supervising elements, which supervise the distance element to ensure the security and selectivity of protection decisions [3]. The "IAL" means that the magnitude of phase A current must exceed the minimum overcurrent threshold (Z50G1). The "IGL" means that the ground current magnitude must exceed a minimum threshold (e.g., 10% of the nominal value). The "32F" means that the directional supervision element must confirm a forward direction of the fault. The FSA is a logic variable and is associated with fault identification selection (FIDS) logic. The conditions for MBG1F and MCG1F are similar to MAG1F, and the details are not shown in Figure 6.1.

 DOI: 10.1201/9781003629481-6

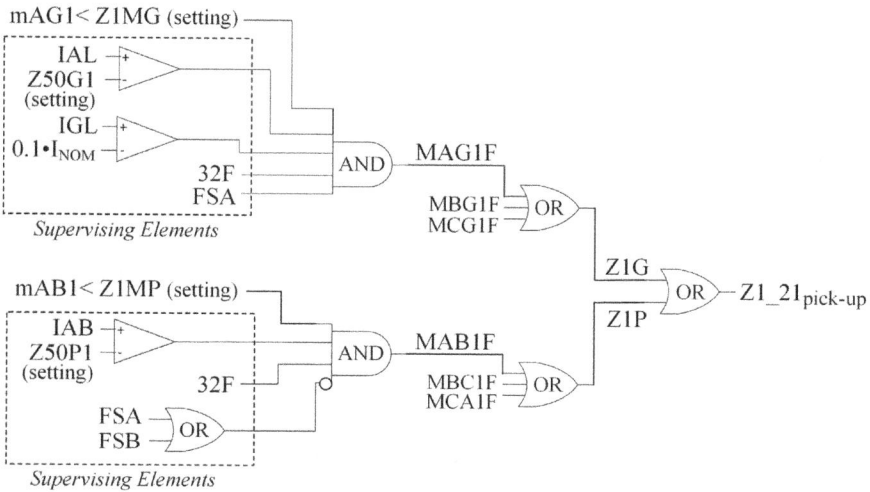

FIGURE 6.1   Logic diagram of a Zone 1 distance element.

The Zone 1 phase-distance (Z1P) element picks up when at least one of the Zone 1 Mho phase-distance elements (e.g., MAB1F) picks up. The "MAB1F" means Zone 1 mho phase-distance element for AB faults, in which the letter "F" represents forward direction. This element picks up only if all the conditions on the left side of the AND gate are true. The "mAB1 < Z1MP" is the main condition, which represents that the impedance calculated by the relay is within the predefined Zone 1 setting. All other conditions below the main condition are supervising elements. The conditions for MBC1F and MCA1F are similar to MAB1F, and the details are not shown in Figure 6.1.

In this chapter, we will first illustrate how impedances are calculated, which is the main principle of distance protection. For the supervising elements, directional supervision has been illustrated in Chapter 5. The fault identification selection (FIDS) logic will be introduced in this chapter.

6.2   IMPEDANCE CHARACTERISTICS FOR DISTANCE PROTECTION

The main principle of distance protection is to calculate an impedance and compare the calculated impedance to a predefined area. If the calculated impedance falls inside the predefined area, then the "mAG1 < Z1MG" condition is satisfied. Mho circle and quadrilateral area are two typical types of predefined areas used for distance relays, as shown in Figure 6.2.

In Figure 6.2, a mho circle and a quadrilateral shape are plotted on an impedance plane. The horizontal axis represents resistance (R), and the vertical axis

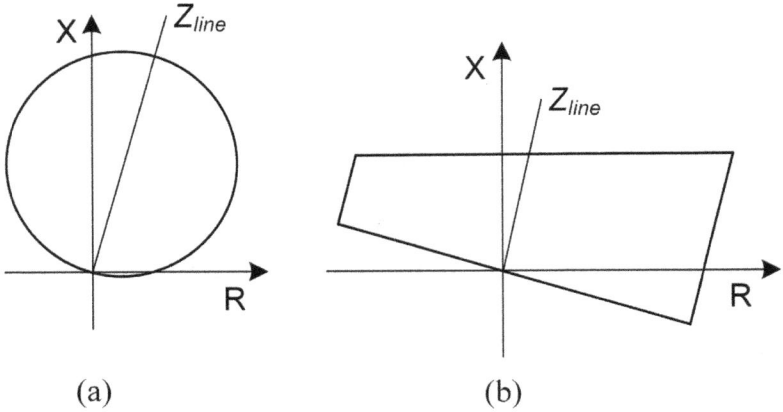

FIGURE 6.2   Example of (a) Mho circle and (b) quadrilateral area.

represents reactance (X). The impedance of the transmission line to be protected is denoted by Z_{line}.

6.3   IMPEDANCE CALCULATION OF MHO CIRCLE

If a Mho circle is used for distance protection, the diameter of the Zone 1 circle is typically set as 80%–85% of the positive-sequence impedance of the transmission line to be protected, as shown in Figure 6.2 (a). Zone 2 circle diameter is typically set as 120%–150% of the transmission line positive-sequence impedance. Zone 2 circle is not shown in Figure 6.2, and the concepts of protection zones will be illustrated in later sections.

The ground-distance element (21G) calculates the AG, BG, and CG effective impedances using Equations (6.1)–(6.3).

$$Z_{AG} = \frac{V_{AG}}{I_A + k_0\left(I_R\right)} \tag{6.1}$$

$$Z_{BG} = \frac{V_{BG}}{I_B + k_0\left(I_R\right)} \tag{6.2}$$

$$Z_{CG} = \frac{V_{CG}}{I_C + k_0\left(I_R\right)} \tag{6.3}$$

where V_{AG}, V_{BG}, and V_{CG} represent phases A, B, and C to ground voltages, respectively. I_A, I_B, I_C represent phase A, B, and C line currents, respectively. I_R represents ground current, also known as residual current. Sometimes I_R is directly measured by a CT. If not directly measured, I_R can be calculated using Equation (6.4).

$$I_R = I_A + I_B + I_C \tag{6.4}$$

The parameter k_0 is calculated using Equation (6.5).

$$k_0 = \frac{Z_0 - Z_1}{3Z_1} \tag{6.5}$$

where Z_0 and Z_1 are the zero-sequence and positive-sequence line impedances, respectively.

The phase-distance element (21P) calculates the AB, BC, and CA effective impedances using Equations (6.6)–(6.8).

$$Z_{AB} = \frac{V_{AG} - V_{BG}}{I_A - I_B} \tag{6.6}$$

$$Z_{BC} = \frac{V_{BG} - V_{CG}}{I_B - I_C} \tag{6.7}$$

$$Z_{CA} = \frac{V_{CG} - V_{AG}}{I_C - I_A} \tag{6.8}$$

It should be noted that the voltage and current phasors are updated sample by sample in a relay, as illustrated in Chapter 3.5. Therefore, the calculated impedances are also updated sample by sample. An example of calculated impedance trajectories during an AG fault event is shown in Figure 6.3.

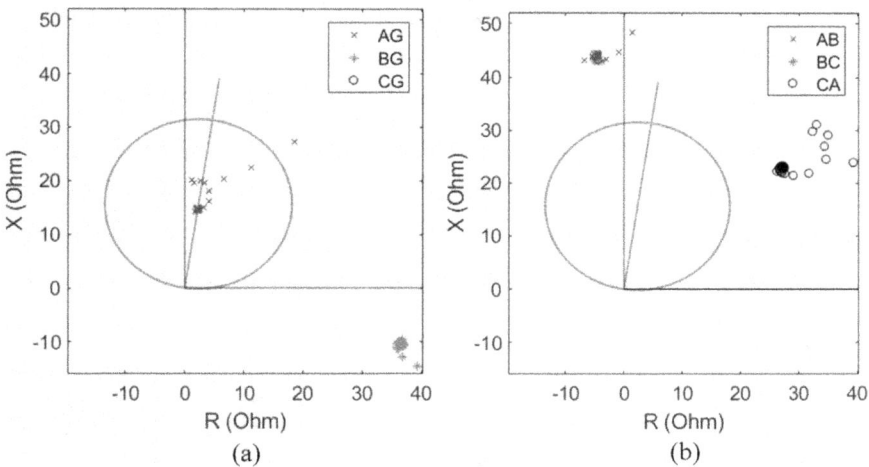

FIGURE 6.3 Impedances calculated by (a) ground-distance elements and (b) phase-distance elements.

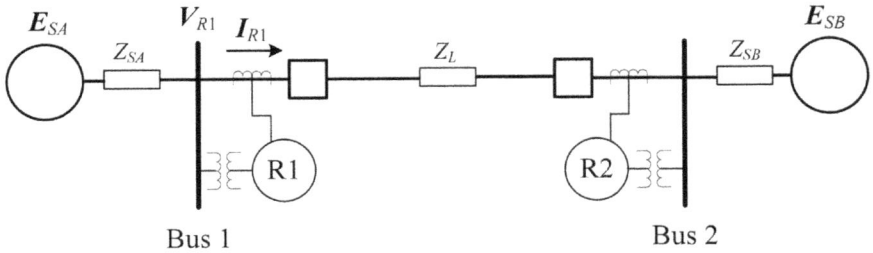

FIGURE 6.4   One-line diagram for Example 6.1.

In Figure 6.3, the simulated fault event corresponds to a phase-A-to-ground (AG) fault occurring on a transmission line. The total length of the transmission line is 80 km, and the fault occurs approximately 30 km from one end of the line. As shown in Figure 6.3, the AG effective impedance converges inside the Mho circle while other effective impedances (e.g., BG and AB) stay outside the Mho circle.

Example 6.1

The one-line diagram of a power system is shown in Figure 6.4. The following parameters are given as CT and VT secondary quantities: $E_{SA}=70\angle 0°\,\text{V}, E_{SB}=70\angle-30°\,\text{V}$. The positive-sequence impedances are $Z_{SA1}=1.5\angle 87°\,\Omega, Z_{SB1}=0.8\angle 83°\,\Omega$, and $Z_{L1}=5\angle 82°\,\Omega$. Current transformer ratio, CTR = 1200/5. Voltage transformer ratio, VTR = 132.8 kV/70 V (line to neutral).

(a) Calculate the source voltages, line and source impedances, and line current referred to the primary side. Also, find the line current in secondary Amps.
(b) For the conditions in part (a), calculate the effective secondary impedances measured by the AG and AB elements of relay R1.
(c) If a three-phase fault occurs at 30% of the location from Bus 1 to Bus 2, calculate the current seen at relay R1 in primary and secondary quantities. You may ignore load flow and assume $E_{SB}=70\angle 0°\,\text{V}$ in this case.
(d) For the conditions in part (c), calculate the effective secondary impedances measured by the AG and AB elements of relay R1.

Solution:

(a) By using the equation for impedance conversion provided in Chapter 2, we can obtain the source A positive-sequence impedance referring to the primary side $Z_{SA1\_prim}=Z_{SA1}\cdot\text{VTR/CTR}=11.86\angle 87°\,\Omega$.
Similarly, we can obtain $Z_{SB1\_prim}=Z_{SB1}\cdot\text{VTR/CTR}=6.32\angle 83°\,\Omega$.
$Z_{L1\_prim}=Z_{L1}\cdot\text{VTR/CTR}=39.52\angle 82°\,\Omega$.

$$E_{SA\_prim} = E_{SA} \cdot \text{VTR} = 132.8\angle 0° \text{V}.$$

$$E_{SB\_prim} = E_{SB} \cdot \text{VTR} = 132.8\angle -30° \text{V}.$$

$$I_{R1\_prim\_A} = \left(E_{SA\_prim} - E_{SB\_prim}\right) / \left(Z_{SA1\_prim} + Z_{SB1\_prim} + Z_{L1\_prim}\right)$$

$$= 1191.99\angle -8.14° \text{A}.$$

$$I_{R1\_sec\_A} = 4.97\angle -8.14° \text{A}. \qquad I_{R1\_sec\_B} = 4.97\angle -128.14° \text{A}.$$

$$V_{R1\_prim\_AG} = 130.81\angle -6.09° \text{V}. \qquad V_{R1\_prim\_BG} = 130.81\angle -126.09° \text{V}.$$

$$V_{R1\_sec\_AG} = 68.95\angle -6.09° \text{V}. \qquad V_{R1\_sec\_BG} = 68.95\angle -126.09° \text{V}.$$

(b) The effective secondary impedances measured by the AG and AB elements of relay R1 are calculated as below.

$$Z_{R1\_sec\_AG} = V_{R1\_sec\_AG} / I_{R1\_sec\_A} = 13.87\angle 2.05° \Omega.$$

Zero-sequence current is not included in the calculation because this is a three-phase balanced case.

$$Z_{R1\_sec\_AB} = \left(V_{R1\_sec\_AG} - V_{R1\_sec\_BG}\right) / \left(I_{R1\_sec\_A} - I_{R1\_sec\_B}\right) = 13.87\angle 2.05° \Omega.$$

(c) When a three-phase fault occurs, $Z_{left1} = Z_{SA1} + 0.3Z_{L1}, Z_{right1} = Z_{SB1} + 0.7Z_{L1}$.

The Thevenin equivalent impedance is $1/Z_{thev1} = \left(1/Z_{left1} + 1/Z_{right1}\right)$.

Therefore, $Z_{thev1} = 0.2 + 1.76j\Omega$.

The total secondary fault current is $I_{f3ph\_sec\_A} = E_{SA} / Z_{thev1} = 39.63\angle -83.55° \text{A}$.

The Fault current (phase A) at relay R1, $I_{f3ph\_R1\_sec\_A} = I_{f3ph\_sec\_A} \cdot Z_{right1} / \left(Z_{left1} + Z_{right1}\right) = 23.36\angle -84.5° \text{A}$.

The Fault current (phase B) at relay R1, $I_{f3ph\_R1\_sec\_B} = 23.36\angle -204.5° \text{A}$.

The secondary voltage (phase A) at relay R1, $V_{f3ph\_R1\_sec\_AG} = 35.03\angle -2.5° \text{V}$.

The secondary voltage (phase B) at relay R1, $V_{f3ph\_R1\_sec\_BG} = 35.03\angle -122.5° \text{V}$.

(d) The effective secondary impedances measured by the AG and AB elements of relay R1 are calculated as below.

$$Z_{f3\ ph\_R1\_sec\_AG} = V_{f3ph\_R1\_sec\_AG} / I_{f3ph\_R1\_sec\_A} = 1.5\angle 82° \Omega.$$

$$Z_{f3ph\_R1\_sec\_AB} = 1.5\angle 82° \Omega.$$

The effective secondary impedances calculated by the AG and AB elements of relay R1 under the load condition (part (b)) are plotted as point M in Figure 6.5.

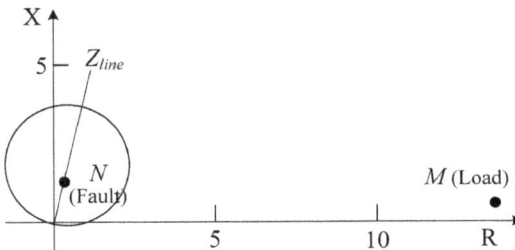

FIGURE 6.5 Impedance plot for Example 6.1.

The effective secondary impedances calculated by the AG and AB elements of relay R1 under the three-phase fault condition (part (d)) are plotted as point N in Figure 6.5. Although the AG and AB impedances are both inside the Mho circle under the three-phase fault condition, only the AB element will pick up. This is because the ground current magnitude does not meet the minimum threshold and thus the AG element is not enabled.

6.4 VOLTAGE POLARIZATION

In Section 6.3, the effective impedances are calculated using the voltages of the same phase(s). For example, the AG effective impedance is calculated using the phase-A-to-ground voltage. This method is called self-polarizing. The self-polarizing method could face challenges when the voltage magnitude is much smaller than its nominal value, which is common for faults occurring close to the relay location and with minor fault resistance. When the voltage magnitude is not sufficient, the accuracy of the calculated impedance is significantly impacted. To overcome the challenges, a relay can use other voltage quantities whose values are more stable during fault events. Such voltages are called polarized voltages, in which "polarized" means "referenced". In this section, we will introduce two commonly used voltage polarization methods.

6.4.1 MEMORY POLARIZING

Memory polarizing is a method of using a buffer to store some previous samples and use these previously stored samples for voltage phasor calculation. Because of the buffer, the calculated voltage phasor magnitude will not decrease as drastically as the situation of using a self-polarizing method, which provides a more stable voltage quantity for impedance calculation. A comparison of voltage phasors calculated by self-polarizing and memory-polarizing methods is shown in Figure 6.6.

In Figure 6.6, the simulated fault event corresponds to an AG fault occurring on a transmission line whose total length is 80 km. The location of the fault is approximately 2 km from one end of the line, which is very close to the relay location. The fault event occurs at approximately 50 ms. The phase-A-to-ground voltage (vAG) magnitude decreases sharply right after the fault occurrence, which can be seen from the voltage phasor magnitude calculated by self-polarizing method ($|VpolAG|_{Self}$). In contrast, the voltage phasor magnitude calculated by memory-polarizing method ($|VpolAG|_{memory}$) stays relatively stable.

A typical memory filter used for calculating the memory-polarized voltage has been illustrated in reference [4]. The polarized voltage for AG element (VpolAG) can be calculated using Equation (6.9).

$$VpolAG[k] = \frac{1}{W}V1[k] - \left(\frac{W-1}{W}\right)VpolAG\left[k - \frac{RS}{2}\right] \qquad (6.9)$$

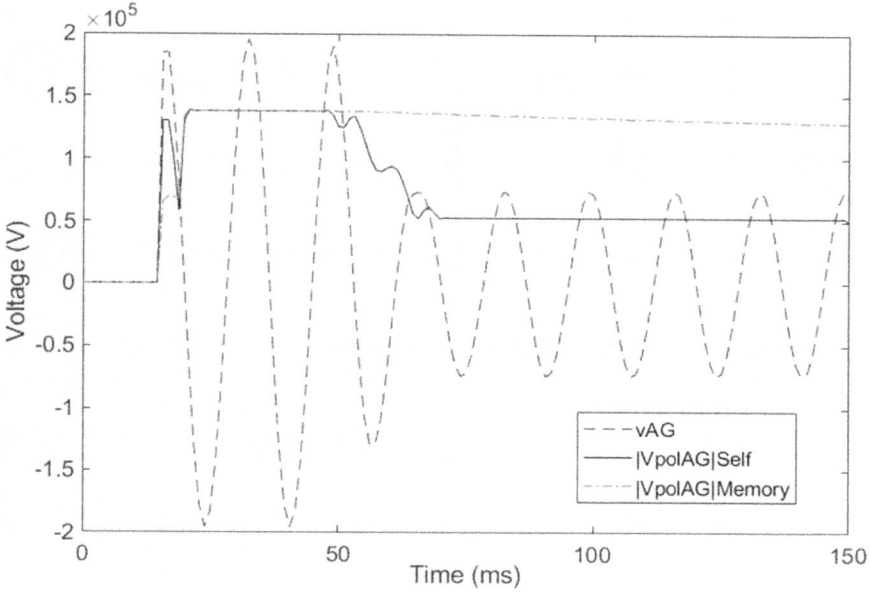

FIGURE 6.6 Comparison between self-polarized and memory-polarized voltages.

In Equation (6.9), the index variable k takes the value from $RS/2$ to the total length of the voltage waveform. RS represents the number of samples per cycle. Thus, the index $(k - RS/2)$ represents a sample that is half a cycle prior to the kth sample. W is an integer number parameter. Selection of W depends on application needs. A smaller W will make VpolAG to change faster (less stable but more sensitive), and a larger W will make VpolAG to change slower (more stable but less sensitive). A typical selection of W could be $W = RS$. V1 represents the positive-sequence voltage, which is calculated from phase A, B, and C voltages.

The polarized voltages for BG and CG elements can be calculated using Equations (6.10) and (6.11), respectively. The parameter a is a constant complex number and $a = 1\angle 120°$.

$$VpolBG[k] = a^2 \, VpolAG[k] \tag{6.10}$$

$$VpolCG[k] = a \, VpolAG[k] \tag{6.11}$$

The polarized voltage for AB element can be calculated using Equation (6.12). The polarized voltages for BC and CA elements can be calculated similarly.

$$VpolAB[k] = VpolAG[k] - VpolBG[k] \tag{6.12}$$

6.4.2 Cross Polarizing

Cross-polarizing is a method of using other uninvolved phases to represent a voltage. For example, the phase-A-to-ground voltage $V_{AG} = -\left(V_{BG} + V_{CG}\right)$ under balanced pre-fault conditions. The polarized voltage for the AG element can be calculated using Equation (6.13), in which VBG and VCG represent phase-B-to-ground and phase-C-to-ground voltages.

$$VpolAG[k] = -\left(VBG[k] + VCG[k]\right) \tag{6.13}$$

The polarized voltages for other elements (e.g., BG, CG, AB, BC, and CA) using the cross-polarizing method are summarized in Table 6.1.

A comparison of voltage phasors calculated by self-polarizing and cross-polarizing methods is shown in Figure 6.7.

TABLE 6.1
Summary of Cross-Polarized Voltages

BG	CG	AB	BC	CA
$-\left(VAG[k] + VCG[k]\right)$	$-\left(VAG[k] + VBG[k]\right)$	$(-\sqrt{3}j).$ $VCG[k]$	$(-\sqrt{3}j).$ $VAG[k]$	$(-\sqrt{3}j).$ $VBG[k]$

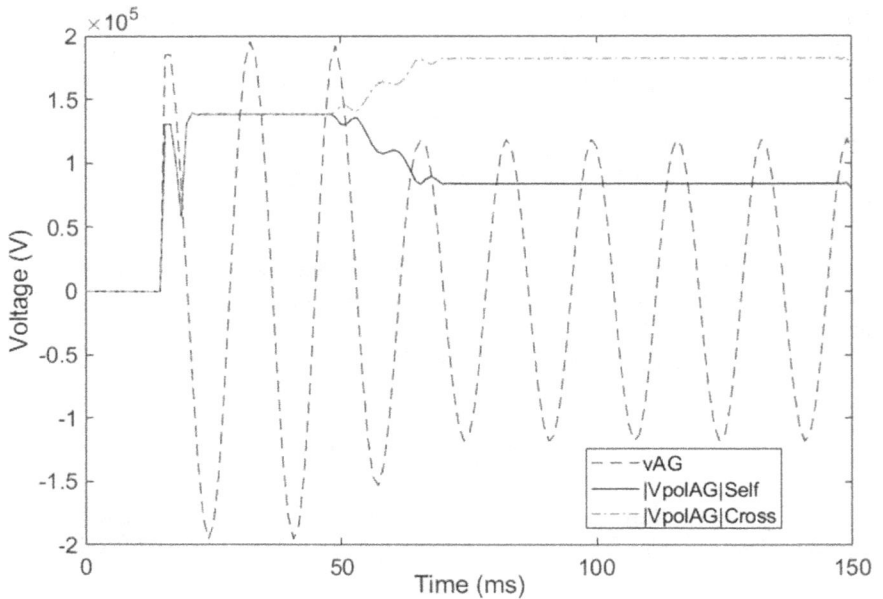

FIGURE 6.7 Comparison between self-polarized and cross-polarized voltages.

In Figure 6.7, the simulated fault event corresponds to an AG fault occurring on a transmission line whose total length is 80 km. The location of the fault is approximately 5 km from one end of the line, which is close to the relay location. The fault event occurs at approximately 50 ms. The cross-polarized voltage magnitude increased after the fault occurrence because the phase-B-to-ground and phase-C-to-ground voltage (not shown in Figure 6.7) magnitudes increased slightly. Another reason is that the angle difference between the phase-B-to-ground and phase-C-to-ground voltages decreased from approximately 120° before fault occurrence to approximately 105° after fault occurrence.

It should be noted that cross-polarization is unreliable for three-phase faults occurring close to the relay because the voltage of each phase is not sufficient. In such situations, memory polarization is preferred.

6.5 MHO CIRCLE EXPANSION

The Mho circle illustrated in Section 6.3 is a static circle passing through the origin point (0, 0), which is only true when a self-polarized voltage is used. When a memory-polarized or cross-polarized voltage is used, a phenomenon called Mho circle expansion will be observed, as shown in Figure 6.8. Z_r represents the impedance reach and Z_s represents an impedance quantity that approximately equals to source impedance.

With expansion, the effective size of a Mho circle increases during a fault event due to the use of a polarized voltage that dynamically adjusts the circle's

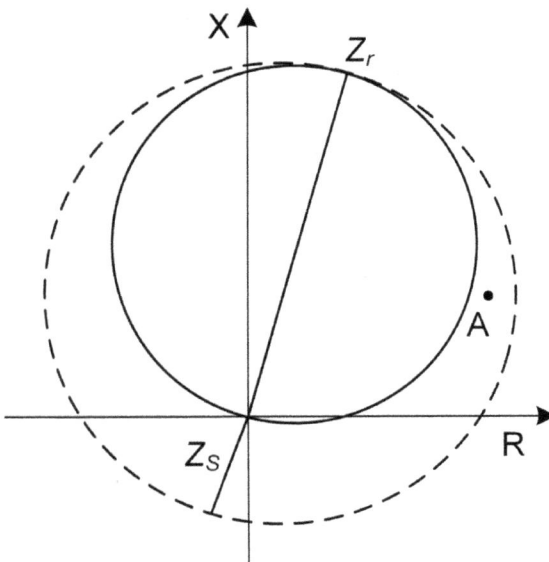

FIGURE 6.8   Brief representation of Mho circle expansion.

reach. This expansion provides better coverage for faults with high fault resistance. For example, if a fault with high fault resistance occurs and the calculated impedance corresponds to point A in Figure 6.8, the distance element would not pick up with the static Mho circle. With the expanded Mho circle, the distance element would pick up.

It should be noted that Mho circle expansion is a dynamic process. The polarized voltage magnitude is changing over time during a fault event, which causes the Mho circle to change its size dynamically. The steady-state resting profile of the dynamic circle depends on multiple factors, such as voltage polarization method, power system parameters, and fault types. Detailed illustrations have been provided in references [5–9].

6.6 CONVERTING AN IMPEDANCE TO A REAL NUMBER

The impedances calculated using Equations (6.1)–(6.3) and (6.6)–(6.8) are complex numbers. A relay has to perform geometric computation to check whether or not a calculated impedance is inside a Mho circle. Such geometric computation is not convenient, especially with the effect of Mho circle expansion. If an impedance can be equivalently converted to a real number and then compared to a threshold, the computational burden can be reduced. Reference [10] provides an algorithm to convert an impedance to a real number M.

The M value for an AG element can be calculated using Equation (6.14), in which the Z_{1L} represents positive-sequence impedance of the transmission line. The V_{AG} represents phase-A-to-ground voltage, V_{polAG} represents the polarized voltage for AG element, and I_A represents phase A current. The bar above a complex number means calculating the conjugate. This equation takes Mho circle expansion effect into account by using the polarized voltage.

$$M_{AG} = m|Z_{1L}| = \frac{Re\left(V_{AG}\,\overline{V_{polAG}}\right)}{Re\left(1\angle Z_{1L}\left(I_A + k_0\,I_R\right)\overline{V_{polAG}}\right)} \tag{6.14}$$

The M values for BG and CG elements can be calculated similarly by replacing V_{AG}, V_{polAG}, and I_A with appropriate quantities.

The M value for an AB element can be calculated using Equation (6.15), in which the V_{polAB} represents polarized voltage for AB element. The M values for BC and CA elements can be calculated similarly.

$$M_{AB} = m|Z_{1L}| = \frac{Re\left((V_{AG} - V_{BG})\overline{V_{polAB}}\right)}{Re\left(1\angle Z_{1L}\left(I_A - I_B\right)\overline{V_{polAB}}\right)} \tag{6.15}$$

The calculated M value is then compared to the magnitude of Zone 1 impedance reach $|Z_{1r}|$. If this M value is between 0 and Z_{1r}, it confirms that the corresponding impedance is inside Zone 1 Mho circle, which also means that the "mAG1 < Z1MG (setting)" or "mAB1 < Z1MP (setting)" condition shown in Figure 6.1 is satisfied. If Zone 1 reach is set as 80% of the transmission line, then $|Z_{1r}| = 0.8|Z_{1L}|$.

6.7 QUADRILATERAL CHARACTERISTICS

The Mho characteristic illustrated in previous sections has been widely used in distance relays for electric power systems. However, the Mho characteristic could face challenges for faults with high fault resistance, especially for high-resistance faults that are close to one end of the transmission line. A quadrilateral element provides better fault resistance coverage compared to a standard Mho element, particularly for short transmission lines with high fault resistance, offering better protection against phase-to-ground faults. A quadrilateral characteristic allows for independent adjustments of the reactance (X) and resistance (R) reach, enabling more precise protection settings.

An example of a quadrilateral characteristic is shown in Figure 6.9. The upper boundary is defined by the reactive reach setting, which is typically configured as 65%–75% of line reactance. The right boundary is defined by the resistive reach setting, which is typically configured as 2–8 times the reactive reach setting, depending on specific situations. The right-side resistive reach setting R_{rset} may differ from the left-side resistive reach setting R_{lset}. A load encroachment blinder is sometimes added on the resistive reach setting boundary because load conditions could possibly go across the boundary. The bottom boundary is defined by a directional element. Detailed quantitative illustrations of quadrilateral characteristics are provided in references [7, 10].

It should be noted that some relay vendors may use different characteristics and terminologies when describing a quadrilateral characteristic. For example,

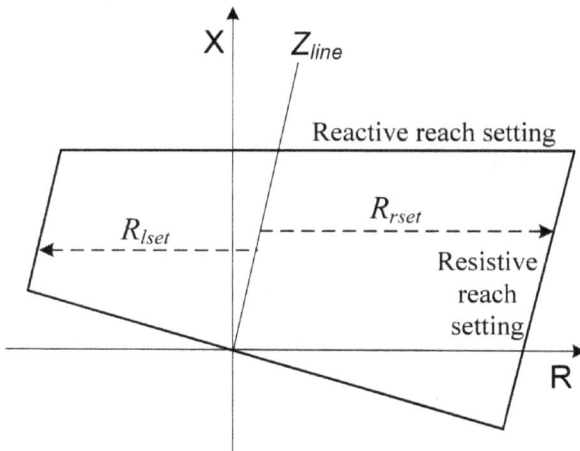

FIGURE 6.9 Example of a quadrilateral characteristic.

some relay vendors use the terminology polygonal [11] instead of quadrilateral, and the characteristic shape is also slightly different from the one shown in Figure 6.9.

Implementing a quadrilateral characteristic can be more complicated than a Mho characteristic due to the need for additional settings to define the quadrilateral boundary. Although quadrilateral characteristics have advantages for short lines with high fault resistance, Mho characteristics may still be preferred for long transmission lines where fault resistance is relatively lower. A Mho element generally offers better immunity to power swings and is less affected by heavy loads compared to a quadrilateral element, making it more suitable for protecting long transmission lines where power flow variations are common.

6.8   FAULT IDENTIFICATION SELECTION LOGIC

The fault identification selection (FIDS) logic has been briefly mentioned in Section 6.1. It is one of the supervising elements that help to improve selectivity in power protection systems. FIDS is used in protective relays to identify the specific phase(s) involved in a fault event and determine which element should be activated. For example, when a phase-A-to-B-to-ground (ABG) fault occurs on a transmission line, it is preferred to block the AG element and let the AB element pick up to avoid overreaching. Sometimes a close-in AG fault could cause the AB and CA effective impedances to enter the Mho circle. In such a case, it is preferred to only enable the AG element.

The FIDS logic is achieved by comparing the angle difference between the zero-sequence current and the negative-sequence current. Detailed illustration has been provided in [12]. Three logic variables, FSA, FSB, and FSC, are used in the FIDS algorithm. The values of the three logic variables can be 0 or 1. The relationship between the logic variable values and the angle difference is summarized in Table 6.2. The angle difference between the zero-sequence current and the negative-sequence current is denoted by $\Delta\theta$.

As shown in Figure 6.1, the AG element will pick up when FSA equals 1 and other conditions are satisfied. The AB element will pick up when both FSA and FSB equal 0 and other conditions are satisfied.

TABLE 6.2
Summary of FIDS Logic

	FSA	FSB	FSC
$-30° \le \Delta\theta \le 30°$	1	0	0
$90° \le \Delta\theta \le 150°$	0	1	0
$-150° \le \Delta\theta \le -90°$	0	0	1
Other Scenarios	0	0	0

6.9 ZONES OF PROTECTION

As we mentioned earlier in this chapter, Zone 1 typically covers 80%–85% of the total length of a transmission line. This is mainly because if a fault occurs close to the remote end of the line, it is difficult to distinguish it from another fault occurring in the beginning section of an adjacent line, especially with fault resistances. To protect the remaining 15%–20% of the line, Zone 2 is used in transmission line relays. Zone 2 is typically set to cover 120%–140% of the line to provide backup protection for some part of the adjacent line. To achieve selectivity, the pickup logic of a Zone 2 element is typically configured to have an intentional delay of 20–25 cycles. Moreover, Zone 3 is used in many transmission line relays to provide farther coverage—typically slightly beyond the entire length of the adjacent line—with a longer delay than Zone 2. A general representation of protection zone coverages is shown in Figure 6.10.

A Zone 2 element is more sensitive than a Zone 1 element. For example, a fault occurring very close to Bus R will not be detected by the Zone 1 element at Bus S, but will be detected by the Zone 2 element at Bus S. A Zone 2 element will not immediately trip when picking up a fault. It will wait for an intentional delay (typically 20–25 cycles) and then trip for the fault. A logic diagram including both Zone 1 and Zone 2 elements is shown in Figure 6.11. For illustration simplicity, Zone 3 is not included in the logic diagram. If Zone 3 is included, it will be represented as another input branch with a longer delay on the left side of the OR gate.

It should be noted that the 120%–140% coverage for Zone 2 is a general guidance. Specific scenarios could be different from the general guidance. For

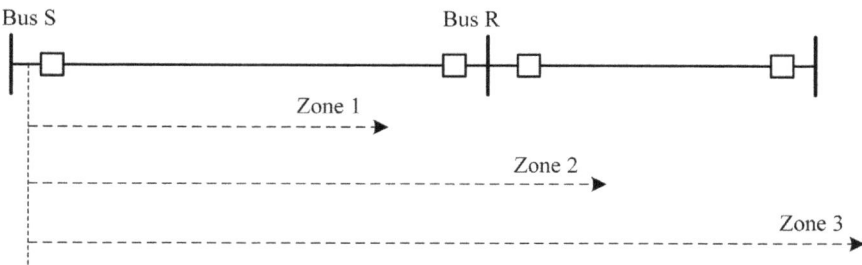

FIGURE 6.10 Coverages of protection zones.

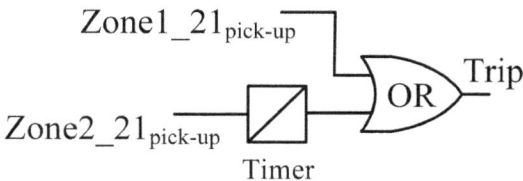

FIGURE 6.11 Trip logic including two zones.

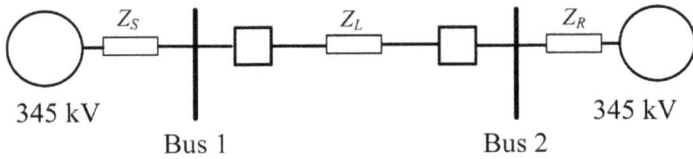

FIGURE 6.12   One-line diagram for Example 6.2.

example, if the adjacent line is much shorter than the main line, the Zone 2 element coverage could be set as 100% of the main line plus 20%–40% of the adjacent line. Detailed fault analysis and coordination studies must be performed when determining the settings.

Example 6.2

The one-line diagram of a power system is shown in Figure 6.12. A distance relay is installed at Bus 1 to protect the transmission line from Bus 1 to Bus 2. Set Zone 1 to protect 80% of the length of the transmission line and Zone 2 to protect 125% of the length of the transmission line.

The positive-, negative-, and zero-sequence impedances of the left-side equivalent voltage source are as follows:
$$Z_{S1} = 2.5\angle85° \ \Omega, Z_{S2} = Z_{S1}, Z_{S0} = 3*Z_{S1}.$$

The positive-, negative-, and zero-sequence impedances of the transmission line are as follows:
$$Z_{L1} = 5\angle85° \ \Omega, Z_{L2} = Z_{L1}, Z_{L0} = 3*Z_{L1}.$$

The positive-, negative-, and zero-sequence impedances of the right-side equivalent voltage source are as follows:
$$Z_{R1} = 2.5\angle85° \ \Omega, Z_{R2} = Z_{R1}, Z_{R0} = 3*Z_{R1}.$$

All the impedance values shown above are in secondary Ohms.
The CT ratio and VT ratio are as follows:
CTR = 800 A/5 A and VTR = 345 kV/120 V (line-to-line)

(a) If a Mho characteristic is used by the distance relay at Bus 1. Assume Bus 1 has a voltage of 1.0 per unit. For a load condition, how much load current (with unity power factor) can flow from Bus 1 to Bus 2 without the Zone 2 element picking up? What if the power factor is 0.8 lagging?
(b) Repeat problem (a) if the power is going from Bus 2 to Bus 1.
(c) With the circuit breaker at Bus 2 open, determine what the following Mho elements (AG, BG, CG, AB, BC, and CA) will calculate for a three-phase fault at 70% of the way down the transmission line from Bus 1 to Bus 2. Repeat with a fault resistance $R_f = 1$ Ohm and with $R_f = 4$ Ohms.

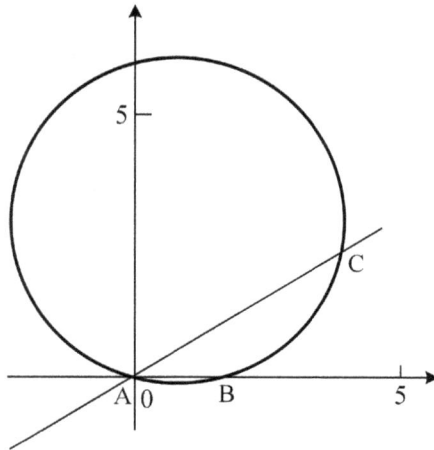

FIGURE 6.13 Mho characteristic for problems (a) and (b) in Example 6.2.

(d) Repeat the calculations for phase-A-to-ground (AG) and phase-B-to-C (BC) faults at 70% of the way down the transmission line from Bus 1 to Bus 2.

(e) Repeat problem (c) if the circuit breaker at Bus 2 is closed.

(f) Repeat problem (d) if the circuit breaker at Bus 2 is closed.

Solution:

To solve problems (a) and (b), let us draw the Zone 2 Mho characteristic of the distance relay, as shown in Figure 6.13. Only the Mho circle for Zone 2 is shown in Figure 6.13. The diameter of the circle is $1.25 * 5 \ \Omega = 6.25 \ \Omega$.

A larger load current magnitude means that an equivalent impedance is more likely to enter the Mho circle. The maximum load current cases correspond to the situations when the calculated load equivalent impedances are on the Zone 2 Mho circle.

(a) If the power is going from Bus 1 to Bus 2 with a unity power factor, the calculated load equivalent impedance corresponds to point B.

Using geometric techniques, we can find that the impedance magnitude corresponding to point B is 0.54 Ω.

$Z_{loadunity} = 0.54 \ \Omega$.

The corresponding secondary load current magnitude is as follows:

$$\frac{V_{secLL}}{\sqrt{3} Z_{loadunity}} = 127.19 \ A$$

The corresponding primary load current magnitude is as follows:

$127.19 * CTR = 20.35 \ kA.$

If the power is going from Bus 1 to Bus 2 with a 0.8 lagging power factor, the calculated load equivalent impedance corresponds to point C. Using geometric techniques, we can find that the impedance magnitude corresponding to point C is 4.17 Ω.

$Z_{loadlag} = 4.17 \ \Omega$.

The corresponding secondary load current magnitude is as follows:

$$\frac{V_{secLL}}{\sqrt{3}Z_{loadlag}} = 16.61 \ A$$

The corresponding primary load current magnitude is as follows:

$16.61 * CTR = 2657.34 \ A$.

(b) If the power is going from Bus 2 to Bus 1 with a unity power factor, the calculated load equivalent impedance corresponds to point A (0, 0).

The impedance magnitude corresponding to point A is 0 Ω.

Since the impedance is 0 Ohm, the corresponding load current is mathematically infinite.

If the power is going from Bus 2 to Bus 1 with a 0.8 lagging power factor, the calculated load equivalent impedance corresponds to point A (0, 0).

Since the impedance is 0 Ohm, the corresponding load current is mathematically infinite.

(c) The circuit breaker at Bus 2 is open.

The positive-, negative-, and zero-sequence impedances to the left of the fault are

$Z_{left1} = Z_{S1} + 0.7 * Z_{L1} = 0.52 + j5.98 \ \Omega$

$Z_{left2} = Z_{S2} + 0.7 * Z_{L2} = 0.52 + j5.98 \ \Omega$

$Z_{left0} = Z_{S0} + 0.7 * Z_{L0} = 1.57 + j17.93 \ \Omega$

The left-side source voltage is 120 V (line-to-line, secondary), which corresponds to 69.28 V (line-to-neutral, secondary). We may set the line-to-neutral voltage angle at 90 degrees.

$V_f = 69.28\angle 90° \ \Omega = j69.28 \ \Omega$.

We initially use $R_f = 0 \ \Omega$ and we will later use $R_f = 1 \ \Omega$ and $R_f = 4 \ \Omega$.

For a three-phase fault, the phase A fault current is as follows:

$$I_{A3ph} = \frac{V_f}{Z_{left1} + R_f} = 11.55\angle 5° \ A$$

The negative- and zero-sequence fault currents are zero.

The phase A voltage at the relay location (Bus 1) is as follows

$V_{A3ph} = V_f - Z_{S1} * I_{A3ph} = 40.41\angle 90° \ V$.

We define a constant at $a = 1\angle 120°$.

The three-phase currents are as follows:

$$\begin{bmatrix} I_{A3ph} \\ I_{B3ph} \\ I_{C3ph} \end{bmatrix} = I_{A3ph} \begin{bmatrix} 1 \\ a^2 \\ a \end{bmatrix} = \begin{bmatrix} 11.55\angle 5° \\ 11.55\angle -115° \\ 11.55\angle 125° \end{bmatrix} A$$

The three-phase voltages at the relay location are as follows:

$$\begin{bmatrix} V_{A3ph} \\ V_{B3ph} \\ V_{C3ph} \end{bmatrix} = V_{A3ph} \begin{bmatrix} 1 \\ a^2 \\ a \end{bmatrix} = \begin{bmatrix} 40.41\angle 90° \\ 40.41\angle -30° \\ 40.41\angle -150° \end{bmatrix} V$$

Using Equations (6.1)–(6.8), we can find the impedances calculated by AG, BG, CG, AB, BC, and CA elements.
When $R_f = 0\ \Omega$,

$$Z_{AG3ph} = Z_{BG3ph} = Z_{CG3ph} = 3.5\ \Omega \text{ with 85 degrees}$$

$ZAB_3ph = ZBC_3ph = ZCA_3ph = 3.5\ \Omega$ with 85 degrees.
When $R_f = 1\ \Omega$,

$$Z_{AG3ph} = Z_{BG3ph} = Z_{CG3ph} = 3.72\ \Omega \text{ with 69.48 degrees}$$

$$Z_{AB3ph} = Z_{BC3ph} = Z_{CA3ph} = 3.72\ \Omega \text{ with 69.48 degrees}$$

When $R_f = 4\ \Omega$,

$$Z_{AG3ph} = Z_{BG3ph} = Z_{CG3ph} = 5.54\ \Omega \text{ with 39 degrees}$$

$$Z_{AB3ph} = Z_{BC3ph} = Z_{CA3ph} = 5.54\ \Omega \text{ with 39 degrees}$$

(d) For a phase-A-to-ground (AG) fault, the positive-, negative-, and zero-sequence fault currents can be calculated as

$$I_{1SLG} = \frac{V_f}{Z_{left1} + Z_{left2} + Z_{left0} + 3R_f}$$

$$I_{2SLG} = I_{0SLG} = I_{1SLG}$$

We initially use $R_f = 0\ \Omega$, and we will later use $R_f = 1\ \Omega$ and $R_f = 4\ \Omega$. The positive-, negative-, and zero-sequence voltages at the relay location can be calculated as

$$V_{1SLG} = V_f - Z_{S1} * I_{1SLG}$$
$$V_{2SLG} = 0 - Z_{S2} * I_{2SLG}$$
$$V_{0SLG} = 0 - Z_{S0} * I_{0SLG}$$

The phase A, B, and C currents are as follows:

$$\begin{bmatrix} I_{ASLG} \\ I_{BSLG} \\ I_{CSLG} \end{bmatrix} = A_{012} \begin{bmatrix} I_{0SLG} \\ I_{1SLG} \\ I_{2SLG} \end{bmatrix} = \begin{bmatrix} 6.93\angle 5° \\ 0 \\ 0 \end{bmatrix} A$$

The matrix A_{012} is a constant matrix that converts sequence quantities to ABC quantities.

$$A_{012} = \begin{bmatrix} 1 & 1 & 1 \\ 1 & a^2 & a \\ 1 & a & a^2 \end{bmatrix}$$

The phase A, B, and C voltages are as follows:

$$\begin{bmatrix} V_{ASLG} \\ V_{BSLG} \\ V_{CSLG} \end{bmatrix} = A_{012} \begin{bmatrix} V_{0SLG} \\ V_{1SLG} \\ V_{2SLG} \end{bmatrix} = \begin{bmatrix} 40.41\angle 90° \\ 75.72\angle -37.59° \\ 75.72\angle -142.41° \end{bmatrix} V$$

Using Equations (6.1)–(6.8), we can find the impedances calculated by AG, BG, CG, AB, BC, and CA elements.

When $R_f = 0\,\Omega$,

Z_{AGSLG} = 3.5 Ω with 85 degrees
Z_{BGSLG} = 16.39 Ω with -42.59 degrees
Z_{CGSLG} = 16.39 Ω with -147.41 degrees
Z_{ABSLG} = 15.2 Ω with 119.72 degrees
Z_{BCSLG} = 120000 Ω with 0 degrees
Z_{CASLG} = 15.21 Ω with 50.28 degrees

When $R_f = 1\,\Omega$,

Z_{AGSLG} = 3.6 Ω with 75.45 degrees
Z_{BGSLG} = 16.78 Ω with -47.59 degrees
Z_{CGSLG} = 16.39 Ω with -152.66 degrees
Z_{ABSLG} = 15.31 Ω with 113.22 degrees
Z_{BCSLG} = 120000 Ω with 0 degrees
Z_{CASLG} = 15.6 Ω with 44 degrees

When $R_f = 4\,\Omega$,

Z_{AGSLG} = 4.41 Ω with 52.19 degrees
Z_{BGSLG} = 18.64 Ω with -60.89 degrees
Z_{CGSLG} = 17.2 Ω with -167.8 degrees
Z_{ABSLG} = 16.74 Ω with 95.27 degrees
Z_{BCSLG} = 120000 Ω with 0 degrees
Z_{CASLG} = 17.75 Ω with 27.64 degrees

Now, we look at a line-to-line (LL) fault. For a phase-B-to-C fault, the positive-, negative-, and zero-sequence fault currents can be calculated as follows:

$$I_{1LL} = \frac{V_f}{Z_{left1} + Z_{left2} + R_f}$$

$$I_{2LL} = -I_{1LL}$$
$$I_{0LL} = 0$$

We initially use $R_f = 0\,\Omega$ and We will later use $R_f = 1\,\Omega$ and $R_f = 4\,\Omega$.

The positive-, negative-, and zero-sequence voltages at the relay location can be calculated as follows:

$$V_{1LL} = V_f - Z_{S1} * I_{1LL}.$$
$$V_{2LL} = 0 - Z_{S2} * I_{2LL}.$$
$$V_{0LL} = 0 - Z_{S0} * I_{0LL}.$$

The phase A, B, and C currents are as follows:

$$\begin{bmatrix} I_{ALL} \\ I_{BLL} \\ I_{CLL} \end{bmatrix} = A_{012} \begin{bmatrix} I_{0LL} \\ I_{1LL} \\ I_{2LL} \end{bmatrix} = \begin{bmatrix} 0\angle 0° \\ 10\angle -85° \\ 10\angle 95° \end{bmatrix} A$$

The matrix A_{012} is a constant matrix that converts sequence quantities to ABC quantities.

$$A_{012} = \begin{bmatrix} 1 & 1 & 1 \\ 1 & a^2 & a \\ 1 & a & a^2 \end{bmatrix}$$

The phase A, B, and C voltages are as follows:

$$\begin{bmatrix} V_{ALL} \\ V_{BLL} \\ V_{CLL} \end{bmatrix} = A_{012} \begin{bmatrix} V_{0LL} \\ V_{1LL} \\ V_{2LL} \end{bmatrix} = \begin{bmatrix} 69.28\angle 90° \\ 49.24\angle -44.7° \\ 49.24\angle -135.3° \end{bmatrix} V$$

Using Equations (6.1)–(6.8), we can find the impedances calculated by AG, BG, CG, AB, BC, and CA elements.

When $R_f = 0 \, \Omega$,

$Z_{AGLL} = 2.4*10^{16} \, \Omega$ with 90 degrees

$Z_{BGLL} = 4.92 \, \Omega$ with 40.3 degrees

$Z_{CGLL} = 4.92 \, \Omega$ with 129.7 degrees

$Z_{ABLL} = 10.97 \, \Omega$ with 13.62 degrees

$Z_{BCLL} = 3.5 \, \Omega$ with 85 degrees

$Z_{CALL} = 10.97 \, \Omega$ with 156.39 degrees

When $R_f = 1 \, \Omega$,

$Z_{AGLL} = 2.4*10^{16} \, \Omega$ with 90 degrees

$Z_{BGLL} = 5.15 \, \Omega$ with 34.23 degrees

$Z_{CGLL} = 4.86 \, \Omega$ with 122.98 degrees

$Z_{ABLL} = 11.29 \, \Omega$ with 8.74 degrees

$Z_{BCLL} = 3.58 \, \Omega$ with 77 degrees

$Z_{CALL} = 10.9 \, \Omega$ with 151.16 degrees

When $R_f = 4 \, \Omega$,

$Z_{AGLL} = 2.4*10^{16} \, \Omega$ with 90 degrees

$Z_{BGLL} = 6.1 \, \Omega$ with 19.43 degrees

$Z_{CGLL} = 5.07 \, \Omega$ with 103.05 degrees

$Z_{ABLL} = 12.69 \, \Omega$ with -3.99 degrees

$Z_{BCLL} = 4.18 \, \Omega$ with 56.53 degrees

$Z_{CALL} = 11.25 \, \Omega$ with 135.69 degrees

(e) The circuit breaker at Bus 2 is closed.

The positive-, negative-, and zero-sequence impedances to the left of the fault are as follows:

$Z_{left1} = Z_{S1} + 0.7*Z_{L1} = 0.52 + j5.98 \, \Omega$

$Z_{left2} = Z_{S2} + 0.7*Z_{L2} = 0.52 + j5.98 \, \Omega$

$Z_{left0} = Z_{S0} + 0.7*Z_{L0} = 1.57 + j17.93 \, \Omega$

The positive-, negative-, and zero-sequence impedances to the right of the fault are as follows

$Z_{right1} = Z_{R1} + 0.3*Z_{L1} = 0.35 + j3.98 \, \Omega$

$$Z_{right2} = Z_{R2} + 0.3 * Z_{L2} = 0.35 + j3.98\ \Omega$$

$$Z_{right0} = Z_{R0} + 0.3 * Z_{L0} = 1.05 + j11.95\ \Omega$$

The Thevenin equivalent positive-, negative-, and zero-sequence imped-ances are as follows:

$$Z_{Thev1} = \left(\frac{1}{Z_{left1}} + \frac{1}{Z_{right1}} \right)^{-1} = 0.21 + j2.39\ \Omega$$

$$Z_{Thev2} = \left(\frac{1}{Z_{left2}} + \frac{1}{Z_{right2}} \right)^{-1} = 0.21 + j2.39\ \Omega$$

$$Z_{Thev0} = \left(\frac{1}{Z_{left0}} + \frac{1}{Z_{right0}} \right)^{-1} = 0.63 + j7.17\ \Omega$$

$$V_f = 69.28\angle 90°\ \Omega = j69.28\ \Omega.$$

We initially use $R_f = 0\ \Omega$ and later we will use $R_f = 1\ \Omega$ and $R_f = 4\ \Omega$. For a three-phase fault, the total phase A fault current is as follows:

$$I_{A3phtotal} = \frac{V_f}{Z_{Thev1} + R_f}$$

The negative- and zero-sequence fault currents are zero.
The left-side phase A fault current is as follows:

$$I_{A3phleft} = I_{A3phtotal} \frac{Z_{right1}}{Z_{left1} + Z_{right1}}$$

The phase A voltage at the relay location (Bus 1) is as follows:

$$V_{A3ph} = V_f - Z_{S1} * I_{A3phleft} = 40.41\angle 90°\ V.$$

We define a constant as $a = 1\angle 120°$.
The three-phase currents are as follows:

$$\begin{bmatrix} I_{A3ph} \\ I_{B3ph} \\ I_{C3ph} \end{bmatrix} = I_{A3phleft} \begin{bmatrix} 1 \\ a^2 \\ a \end{bmatrix} = \begin{bmatrix} 11.55\angle 5° \\ 11.55\angle -115° \\ 11.55\angle 125° \end{bmatrix} A$$

The three-phase voltages at the relay location are as follows:

$$\begin{bmatrix} V_{A3ph} \\ V_{B3ph} \\ V_{C3ph} \end{bmatrix} = V_{A3ph} \begin{bmatrix} 1 \\ a^2 \\ a \end{bmatrix} = \begin{bmatrix} 40.41\angle 90° \\ 40.41\angle -30° \\ 40.41\angle -150° \end{bmatrix} V$$

We can see that when $R_f = 0\ \Omega$, the results are the same as the ones in problem (c).
Using Equations (6.1)–(6.8), we can find the impedances calculated by AG, BG, CG, AB, BC, and CA elements.
When $R_f = 0\ \Omega$,

$$Z_{AG3ph} = Z_{BG3ph} = Z_{CG3ph} = 3.5\ \Omega \text{ with } 85 \text{ degrees}$$

$$Z_{AB3ph} = Z_{BC3ph} = Z_{CA3ph} = 3.5\ \Omega \text{ with } 85 \text{ degrees}$$

When $R_f = 1 \, \Omega$,

$Z_{AG3ph} = Z_{BG3ph} = Z_{CG3ph} = 4.47 \, \Omega$ with 51.18 degrees

$Z_{AB3ph} = Z_{BC3ph} = Z_{CA3ph} = 4.47 \, \Omega$ *with 51.18 degrees*

When $R_f = 4 \, \Omega$,

$Z_{AG3ph} = Z_{BG3ph} = Z_{CG3ph} = 10.88 \, \Omega$ with 18.7 degrees

$Z_{AB3ph} = Z_{BC3ph} = Z_{CA3ph} = 10.88 \, \Omega$ with 18.7 degrees

When R_f is nonzero, the results are different from the ones in problem (c).

(f) The circuit breaker at Bus 2 is closed.

For a phase-A-to-ground (AG) fault, the positive-, negative-, and zero-sequence total currents can be calculated as

$$I_{1SLGtotal} = \frac{V_f}{Z_{Thev1} + Z_{Thev2} + Z_{Thev0} + 3R_f}$$

$$I_{2SLGtotal} = I_{0SLGtotal} = I_{1SLGtotal}$$

We initially use $R_f = 0 \, \Omega$ and later we will use $R_f = 1 \, \Omega$ and $R_f = 4 \, \Omega$. The left-side positive-, negative-, and zero-sequence fault currents are as follows:

$$I_{1SLGleft} = I_{1SLGtotal} \frac{Z_{right1}}{Z_{left1} + Z_{right1}}$$

$$I_{2SLGleft} = I_{2SLGtotal} \frac{Z_{right2}}{Z_{left2} + Z_{right2}}$$

$$I_{0SLGleft} = I_{0SLGtotal} \frac{Z_{right0}}{Z_{left0} + Z_{right0}}$$

The positive-, negative-, and zero-sequence voltages at the relay location (Bus 1) are as follows:

$$V_{1SLG} = V_f - Z_{S1} * I_{1SLGleft}$$

$$V_{2SLG} = \cdot 0 \cdot - Z_{S2} * I_{2SLGleft}$$

$$V_{0SLG} = 0 - Z_{S0} * I_{0SLGleft}$$

The phase A, B, and C currents are as follows:

$$\begin{bmatrix} I_{ASLG} \\ I_{BSLG} \\ I_{CSLG} \end{bmatrix} = A_{012} \begin{bmatrix} I_{0SLGleft} \\ I_{1SLGleft} \\ I_{2SLGleft} \end{bmatrix} = \begin{bmatrix} 6.93 \angle 5° \\ 0 \\ 0 \end{bmatrix} A \quad \text{The matrix } A_{012} \text{ is a constant}$$

matrix that converts sequence quantities to ABC quantities.

$$A_{012} = \begin{bmatrix} 1 & 1 & 1 \\ 1 & a^2 & a \\ 1 & a & a^2 \end{bmatrix}$$

The phase A, B, and C voltages are as follows:

$$\begin{bmatrix} V_{ASLG} \\ V_{BSLG} \\ V_{CSLG} \end{bmatrix} = A_{012} \begin{bmatrix} V_{0SLG} \\ V_{1SLG} \\ V_{2SLG} \end{bmatrix} = \begin{bmatrix} 40.41\angle 90° \\ 75.72\angle -37.59° \\ 75.72\angle -142.41° \end{bmatrix} V$$

We can see that when $R_f = 0\Omega$, the results are the same as the ones in problem (d).

Using Equations (6.1)–(6.8), we can find the impedances calculated by AG, BG, CG, AB, BC, and CA elements.

When $R_f = 0\Omega$,

$Z_{AGSLG} = 3.5\ \Omega$ with 85 degrees

$Z_{BGSLG} = 16.39\ \Omega$ with -42.59 degrees

$Z_{CGSLG} = 16.39\ \Omega$ with -147.41 degrees

$Z_{ABSLG} = 15.2\ \Omega$ with 119.72 degrees

$Z_{BCSLG} = 120000\ \Omega$ with 0 degrees

$Z_{CASLG} = 15.21\ \Omega$ with 50.28 degrees

When $R_f = 1\Omega$,

$Z_{AGSLG} = 3.93\ \Omega$ with 62.63 degrees

$Z_{BGSLG} = 17.6\ \Omega$ with -54.59 degrees

$Z_{CGSLG} = 16.65\ \Omega$ with -160.41 degrees

$Z_{ABSLG} = 15.83\ \Omega$ with 103.85 degrees

$Z_{BCSLG} = 120000\ \Omega$ with 0 degrees

$Z_{CASLG} = 16.5\ \Omega$ with 35.29 degrees

When $R_f = 4\Omega$,

$Z_{AGSLG} = 7.2\ \Omega$ with 28.94 degrees

$Z_{BGSLG} = 24.51\ \Omega$ with -79.26 degrees

$Z_{CGSLG} = 21.71\ \Omega$ with 168.95 degrees

$Z_{ABSLG} = 23.1\ \Omega$ with 71.17 degrees

$Z_{BCSLG} = 120000\ \Omega$ with 0 degrees

$Z_{CASLG} = 24.91\ \Omega$ with 7 degrees

When R_f is nonzero, the results are different from the ones in problem (d). Now we look at a line-to-line (LL) fault. For a phase-B-to-C fault, the positive-, negative-, and zero-sequence total currents can be calculated as

$$I_{1LLtotal} = \frac{V_f}{Z_{Thev1} + Z_{Thev2} + R_f}$$

$$I_{2LLtotal} = -I_{1LLtotal}$$

$$I_{0LLtotal} = 0$$

We initially use $R_f = 0\ \Omega$ and later we will use $R_f = 1\ \Omega$ and $R_f = 4\ \Omega$.

The left-side positive-, negative-, and zero-sequence fault currents are as follows:

$$I_{1LLleft} = I_{1LLtotal} \frac{Z_{right1}}{Z_{left1} + Z_{right1}}$$

$$I_{2LLleft} = I_{2LLtotal} \frac{Z_{right2}}{Z_{left2} + Z_{right2}}$$

$$I_{0LLleft} = I_{0LLtotal} \frac{Z_{right0}}{Z_{left0} + Z_{right0}}$$

The positive-, negative-, and zero-sequence voltages at the relay location can be calculated as follows:

$$V_{1LL} = V_f - Z_{S1} * I_{1LLleft}$$

$$V_{2LL} = 0 \ -Z_{S2} * I_{2LLleft}$$

$$V_{0LL} = 0 \ -Z_{S0} * I_{0LLleft}$$

The phase A, B, and C currents are as follows:

$$\begin{bmatrix} I_{ALL} \\ I_{BLL} \\ I_{CLL} \end{bmatrix} = A_{012} \begin{bmatrix} I_{0LLleft} \\ I_{1LLleft} \\ I_{2LLleft} \end{bmatrix} = \begin{bmatrix} 0\angle 0° \\ 10\angle -85° \\ 10\angle 95° \end{bmatrix} A$$

The matrix A_{012} is a constant matrix that converts sequence quantities to ABC quantities.

$$A_{012} = \begin{bmatrix} 1 & 1 & 1 \\ 1 & a^2 & a \\ 1 & a & a^2 \end{bmatrix}$$

The phase A, B, and C voltages are as follows:

$$\begin{bmatrix} V_{ALL} \\ V_{BLL} \\ V_{CLL} \end{bmatrix} = A_{012} \begin{bmatrix} V_{0LL} \\ V_{1LL} \\ V_{2LL} \end{bmatrix} = \begin{bmatrix} 69.28\angle 90° \\ 49.24\angle -44.7° \\ 49.24\angle -135.3° \end{bmatrix} V$$

When $R_f = 0\ \Omega$, the results are the same as the ones in problem (d). Using Equations (6.1)–(6.8), we can find the impedances calculated by AG, BG, CG, AB, BC, and CA elements.
When $R_f = 0\ \Omega$,

$Z_{AGLL} = 2.4*10^{16}\ \Omega$ with 90 degrees

$Z_{BGLL} = 4.92\ \Omega$ with 40.3 degrees

$Z_{CGLL} = 4.92\ \Omega$ with 129.7 degrees

$Z_{ABLL} = 10.97\ \Omega$ with 13.62 degrees

$Z_{BCLL} = 3.5\ \Omega$ with 85 degrees

$Z_{CALL} = 10.97\ \Omega$ with 156.39 degrees

When $R_f = 1 \, \Omega$,

$Z_{AGLL} = 2.4 * 10^{16} \, \Omega$ with 90 degrees

$Z_{BGLL} = 5.58 \, \Omega$ with 26.2 degrees

$Z_{CGLL} = 4.89 \, \Omega$ with 112.8 degrees

$Z_{ABLL} = 11.92 \, \Omega$ with 2.01 degrees

$Z_{BCLL} = 3.82 \, \Omega$ with 65.96 degrees

$Z_{CALL} = 10.97 \, \Omega$ with 143.31 degrees

When $R_f = 4 \, \Omega$,

$Z_{AGLL} = 2.4 * 10^{16} \, \Omega$ with 90 degrees

$Z_{BGLL} = 8.76 \, \Omega$ with 1.95 degrees

$Z_{CGLL} = 6.93 \, \Omega$ with 74.5 degrees

$Z_{ABLL} = 16.8 \, \Omega$ with -21.21 degrees

$Z_{BCLL} = 6.35 \, \Omega$ with 33.31 degrees

$Z_{CALL} = 14 \, \Omega$ with 111.15 degrees

When R_f is nonzero, the results are different from the ones in problem (d).

6.10 MATLAB IMPLEMENTATION

In this section, we will illustrate the implementation of distance protection functions using MATLAB. The functions are developed for a distance relay located at one end of a transmission line. The transmission line, positive-sequence impedance is $39.59\angle81.56°\, \Omega$. The negative-sequence impedance equals the positive-sequence impedance. The zero-sequence impedance is $101.38\angle68.81°\, \Omega$. In this example, only the Zone 1 protection using Mho characteristic is included. Implementation of quadrilateral characteristic is beyond the scope of this book.

The *Relay1_main.m* is the master file. Running this file will run the entire distance relay program. The MATLAB code of the *Relay1_main.m* file is shown as follows:

```
%Relay1_main.m, the master file of a distance relay
%Begin
clear all;
close all;
clc;
%Read voltage and current data
Relay1_readdata;
%Configure the Relay1 settings
Relay1_setting;
%Process the input currents and voltages through a filter
Relay1_filter;
%Create Phasors and Vpol
Relay1_phasor_vpol;
%Perform calculation for Ground distance elements (21G)
Relay1_21G_calc;
```

```
%Perform calculation for Phase distance elements (21P)
Relay1_21P_calc;
%Run directional elements
Relay1_directional;
%Run the FID logic
Relay1_fid_logic;
%Run the pickup logic for Zone 1 Ground distance elements
(21G)
Relay1_21G_zone1;
%Run the pickup logic for Zone 1 Phase distance elements
(21P)
Relay1_21P_zone1;
%Run the overall pickup logic for Zone 1 distance element
Relay1_21_zone1;
%Output the results
Relay1_plot;
%End
```

The *Relay1_readdata.m* reads the input voltages and currents. In this example, the input data are saved in a MATLAB data file *system1_30_50.mat*. The MATLAB code of the *Relay1_readdata.m* file is shown below. The data are also resampled to 16 samples per cycle.

```
%Relay1_readdata.m, reading the input data
%Begin
CTR = 1.0;
PTR = 1.0;
VIread = load('system1_30_50.mat');
Ia = VIread.iBus02aBus03a;
Ib = VIread.iBus02bBus03b;
Ic = VIread.iBus02cBus03c;
Va = VIread.vBus03a;
Vb = VIread.vBus03b;
Vc = VIread.vBus03c;
time_before = VIread.t;
dt_before = time_before(2)-time_before(1);
RS_before = round( (1*3.0)/(60*dt_before) );
RS_after = 16*3;
IA = resample(Ia, RS_after, RS_before);
IB = resample(Ib, RS_after, RS_before);
IC = resample(Ic, RS_after, RS_before);
VA = resample(Va, RS_after, RS_before);
VB = resample(Vb, RS_after, RS_before);
VC = resample(Vc, RS_after, RS_before);
len = length(IA);
for i=1:len
    IR(i) = IA(i)+IB(i)+IC(i);
end
%End
```

The *Relay1_setting.m* configures the relay settings. Note that the line impedances in this example are different from the ones shown in Chapter 3. We have adjusted the line impedances in this example. Also, we have set Z1MG = 0.8*Z1MAG and Z1MP = Z1MG, which means the Zone 1 reaches for the Mho ground- and phase-distance elements are both set at 80% of the line positive-sequence impedance. The MATLAB code of the *Relay1_setting.m* file is shown as follows:

```
%Relay1_setting.m
%Begin
Z1MAG = 39.5887; %Positive-sequence line impedance magnitude
Z1ANG = 81.56*pi/180; %Positive-sequence line impedance
angle
[xx, yy] = pol2cart(Z1ANG,Z1MAG);
Z1Impedance = xx + 1i*yy;
Z0MAG = 101.3806; %Zero-sequence line impedance magnitude
Z0ANG = 68.81*pi/180; %Zero-sequence line impedance angle
[xx, yy]=pol2cart(Z0ANG,Z0MAG);
Z0Impedance = xx + 1i*yy;
k0 = (Z0Impedance-Z1Impedance)/(3.0*Z1Impedance);
Inom = 5; %5A or 1A
%Settings for directional supervision element
ZFthre = 0.49*Z1MAG;
ZRthre = 0.51*Z1MAG;
a2 = 0.1;
F50Q = 0.3;
R50Q = 0.3;

%Voltage polarization options:
%Option 1:self-polarizing, 2:cross-polarizing
Vpol_option = 1;
%Zone 1 Mho distance reach settings
Z1MG = 0.8*Z1MAG; %Mho Grond distance reach set as 80% of
Z1MAG
Z1MP = Z1MG; %Mho Phase distance reach set as 80% of Z1MAG
%Zone 1 Ground and Phase fault current minimum thresholds
Z50G1 = 0.5;
Z50P1 = 0.5;
%End
```

The *Relay1_filter.m* performs a filter algorithm to voltages and currents. The MATLAB code is shown as follows:

```
%Relay1_filter.m
%Begin
RS = 16;
filter = 1;
RANGE = 4;
a = -0.5 + 0.5*sqrt(3)*1i;
%Filter, if (filter==0), Filter will be skipped
```

```
%{
IA, IB, IC, IR, VA, VB, VC are the
currents/voltages after being processed by the filter
%}
if (filter==1)
    Ir = zeros(len,1);
    Ia = zeros(len,1);
    Ib = zeros(len,1);
    Ic = zeros(len,1);
    Va = zeros(len,1);
    Vb = zeros(len,1);
    Vc = zeros(len,1);
    for i = RS:len
        for k = 1:RS
            Ir(i) = Ir(i)+cos(2*pi*(k-1)/RS)*IR(i-RS+k);
        end
        Ir(i) = Ir(i)*2/RS;
        for k = 1:RS
            Ia(i) = Ia(i)+cos(2*pi*(k-1)/RS)*IA(i-RS+k);
        end
        Ia(i) = Ia(i)*2/RS;
        for k = 1:RS
            Ib(i) = Ib(i)+cos(2*pi*(k-1)/RS)*IB(i-RS+k);
        end
        Ib(i) = Ib(i)*2/RS;
        for k = 1:RS
            Ic(i) = Ic(i)+cos(2*pi*(k-1)/RS)*IC(i-RS+k);
        end
        Ic(i) = Ic(i)*2/RS;
        for k = 1:RS
            Va(i) = Va(i)+cos(2*pi*(k-1)/RS)*VA(i-RS+k);
        end
        Va(i) = Va(i)*2/RS;
        for k = 1:RS
            Vb(i) = Vb(i)+cos(2*pi*(k-1)/RS)*VB(i-RS+k);
        end
        Vb(i) = Vb(i)*2/RS;
        for k = 1:RS
            Vc(i) = Vc(i)+cos(2*pi*(k-1)/RS)*VC(i-RS+k);
        end
        Vc(i) = Vc(i)*2/RS;
    end
    IA = Ia;
    IB = Ib;
    IC = Ic;
    IR = Ir;
    VA = Va;
    VB = Vb;
    VC = Vc;
end
%End
```

The *Relay1_phasor_vpol.m* creates phasors and calculates polarizing voltages for further protection functions. The MATLAB code is shown as follows:

```
%Relay1_phasor_vpol.m
%Begin
IAcpx = zeros(1,len);
IBcpx = zeros(1,len);
ICcpx = zeros(1,len);
IRcpx = zeros(1,len);
VAcpx = zeros(1,len);
VBcpx = zeros(1,len);
VCcpx = zeros(1,len);
IABcpx = zeros(1,len);
IBCcpx = zeros(1,len);
ICAcpx = zeros(1,len);
VABcpx = zeros(1,len);
VBCcpx = zeros(1,len);
VCAcpx = zeros(1,len);
I0 = zeros(1,len);
I1 = zeros(1,len);
I2 = zeros(1,len);
V0 = zeros(1,len);
V1 = zeros(1,len);
V2 = zeros(1,len);
for v = (RS/4+1):len
    IAcpx(v) = ( IA(v)+ 1i*IA(v-RS/4) )/sqrt(2);
    IBcpx(v) = ( IB(v)+ 1i*IB(v-RS/4) )/sqrt(2);
    ICcpx(v) = ( IC(v)+ 1i*IC(v-RS/4) )/sqrt(2);
    IRcpx(v) = ( IR(v)+ 1i*IR(v-RS/4) )/sqrt(2);
    VAcpx(v) = ( VA(v)+ 1i*VA(v-RS/4) )/sqrt(2);
    VBcpx(v) = ( VB(v)+ 1i*VB(v-RS/4) )/sqrt(2);
    VCcpx(v) = ( VC(v)+ 1i*VC(v-RS/4) )/sqrt(2);

    IABcpx(v) = IAcpx(v)-IBcpx(v);
    IBCcpx(v) = IBcpx(v)-ICcpx(v);
    ICAcpx(v) = ICcpx(v)-IAcpx(v);

    VABcpx(v) = VAcpx(v)-VBcpx(v);
    VBCcpx(v) = VBcpx(v)-VCcpx(v);
    VCAcpx(v) = VCcpx(v)-VAcpx(v);

    I0(v) = (IAcpx(v)+IBcpx(v)+ICcpx(v))*(1.0/3);
    I1(v) = (IAcpx(v)+a*IBcpx(v)+a*a*ICcpx(v))*(1.0/3);
    I2(v) = (IAcpx(v)+a*a*IBcpx(v)+a*ICcpx(v))*(1.0/3);

    V0(v) = (VAcpx(v)+VBcpx(v)+VCcpx(v))*(1.0/3);
    V1(v) = (VAcpx(v)+a*VBcpx(v)+a*a*VCcpx(v))*(1.0/3);
    V2(v) = (VAcpx(v)+a*a*VBcpx(v)+a*VCcpx(v))*(1.0/3);
end
```

```
VpolAG = zeros(1,len);
VpolBG = zeros(1,len);
VpolCG = zeros(1,len);
VpolAB = zeros(1,len);
VpolBC = zeros(1,len);
VpolCA = zeros(1,len);
if (Vpol_option == 1) %Option 1, using self-polarizing
    VpolAG = VAcpx;
    VpolBG = VBcpx;
    VpolCG = VCcpx;
    VpolAB = VABcpx;
    VpolBC = VBCcpx;
    VpolCA = VCAcpx;
elseif (Vpol_option == 2) %Option 2, using cross-polarizing
    VpolAG = -(VBcpx + VCcpx);
    VpolBG = -(VAcpx + VCcpx);
    VpolCG = -(VAcpx + VBcpx);
    VpolAB = (-sqrt(3)*1i)*VCcpx;
    VpolBC = (-sqrt(3)*1i)*VAcpx;
    VpolCA = (-sqrt(3)*1i)*VBcpx;
end
%End
```

The *Relay1_21G_calc.m* and *Relay1_21P_calc.m* files calculate the *M* values for ground-distance (21G) and phase-distance (21P) elements. The MATLAB code of the *Relay1_21G_calc.m* and *Relay1_21P_calc.m* files is shown below. The variables UP and DOWN represent a numerator and a denominator, respectively. The purpose of calculating a numerator and a denominator individually is to avoid having a line of code that is too long.

```
%Relay1_21G_calc.m, Impedance calculation for Ground
distance elements (Mho)
%Begin
[xx, yy]=pol2cart(Z1ANG, 1.0);
Z1cpx = xx + 1i*yy;
MAG = zeros(1,len);
MBG = zeros(1,len);
MCG = zeros(1,len);
for i = 1:len
    %Calculate MAG, MBG, MCG
    UP = real(VAcpx(i)*conj(VpolAG(i)));
    DOWN = real(Z1cpx*(IAcpx(i)+k0*IRcpx(i))*conj(VpolA
G(i)))+0.00001;
    MAG(i) = UP/DOWN;

    UP = real(VBcpx(i)*conj(VpolBG(i)));
    DOWN = real(Z1cpx*(IBcpx(i)+k0*IRcpx(i))*conj(VpolB
G(i)))+0.00001;
    MBG(i) = UP/DOWN;
```

```
    UP = real(VCcpx(i)*conj(VpolCG(i)));
    DOWN = real(Z1cpx*(ICcpx(i)+k0*IRcpx(i))*conj(VpolC
G(i)))+0.00001;
    MCG(i) = UP/DOWN;
end
%End
%Relay1_21P_calc.m, Impedance calculation for Phase distance
elements (Mho)
%Begin
MABD = zeros(1,len);
MBCD = zeros(1,len);
MCAD = zeros(1,len);
MAB = zeros(1,len);
MBC = zeros(1,len);
MCA = zeros(1,len);
Z1ABC = zeros(1,len);
R1ABC = zeros(1,len);
X1ABC = zeros(1,len);
    [xx, yy]=pol2cart(Z1ANG, 1.0);
Z1cpx = xx + 1i*yy;
for i = 1:len
    MABD(i)= real( Z1cpx*(IAcpx(i)-IBcpx(i))*conj(VpolAB(i)));
    MABD(i) = MABD(i) + 0.00001;
    MBCD(i)= real( Z1cpx*(IBcpx(i)-ICcpx(i))*conj(VpolBC(i)));
    MBCD(i) = MBCD(i) + 0.00001;
    MCAD(i)= real( Z1cpx*(ICcpx(i)-IAcpx(i))*conj(VpolCA(i)));
    MCAD(i) = MCAD(i) + 0.00001;

    MAB(i)= real((VAcpx(i)-VBcpx(i))*conj(VpolAB(i)))/MABD(i);
    MBC(i)= real((VBcpx(i)-VCcpx(i))*conj(VpolBC(i)))/MBCD(i);
    MCA(i)= real((VCcpx(i)-VAcpx(i))*conj(VpolCA(i)))/MCAD(i);

    %Positive sequence impedance
    Z1ABC(i) = V1(i)/(I1(i)+0.00001);
    R1ABC(i) = real(Z1ABC(i));
    X1ABC(i) = imag(Z1ABC(i));
end
%End
```

The *Relay1_directional.m* file performs computation for a negative-sequence impedance-based directional supervision function. The MATLAB code is shown below. After running the directional supervision function, if the logic variable F32Q(i) equals 1, it confirms a forward direction. If the logic variable R32Q(i) equals 1, it confirms a reverse direction.

```
%Relay1_directional.m, Directional element
%Begin
Q50F = zeros(1,len);
Q50R = zeros(1,len);
A2 = zeros(1,len);
```

```
Q32E = zeros(1,len);
Z2 = zeros(1,len);
Z0 = zeros(1,len);
F32Q = zeros(1,len);
R32Q = zeros(1,len);
[xx, yy]=pol2cart(Z1ANG, 1.0);
Z1cpx = xx + 1i*yy;
[xx, yy]=pol2cart(Z0ANG, 1.0);
Z0cpx = xx + 1i*yy;
for i = 1:len
    if (abs(3*I2(i))>F50Q)
        Q50F(i) = 1;
    else
        Q50F(i) = 0;
    end

    if (abs(3*I2(i))>R50Q)
        Q50R(i) = 1;
    else
        Q50R(i) = 0;
    end

if (abs(I2(i))/abs(I1(i))>=a2)
        A2(i) = 1;
    else
        A2(i) = 0;
    end

    if (((Q50F(i)==1)||(Q50R(i)==1)) && A2(i)==1)
        Q32E(i)=1;
    else
        Q32E(i)=0;
    end

    UP = real( V2(i)*conj(I2(i)*Z1cpx));
    DOWN = abs(I2(i))*abs(I2(i))+0.00001;
    Z2(i) = UP/DOWN;

    UP = real( V0(i)*conj(I0(i)*Z0cpx));
    DOWN = abs(I0(i))*abs(I0(i))+0.00001;
    Z0(i) = UP/DOWN;

    if (Z2(i)<=ZFthre)&&(Q32E(i)==1)
        F32Q(i)=1;
    else
        F32Q(i)=0;
    end

    if (Z2(i)>=ZRthre)&&(Q32E(i)==1)
        R32Q(i)=1;
```

```
    else
        R32Q(i)=0;
    end
end
%End
```

The *Relay1_fid_logic.m* file includes the FIDS logic computation. The MATLAB code is shown below. After running this function, the values of logic variables FSA, FSB, and FSC will be determined.

```
%Relay1_fid_logic.m, FID Logic
%Begin
Ibeta_max = zeros(1,len);
for i = 1:len
    Ibeta_max(i) = max([abs(IABcpx(i)), abs(IBCcpx(i)),
abs(ICAcpx(i))]);
end
IRB = zeros(1,len);
I02 = zeros(1,len);
SPO = zeros(1,len);
FIDEN = zeros(1,len);
for i = 1:len
    if abs(IRcpx (i))>0.1*Ibeta_max(i)
        IRB(i) = 1;
    else
        IRB(i) = 0;
    end

    if abs(I0(i))>0.1*abs(I2(i))
        I02(i) = 1;
    else
        I02(i) = 0;
end

    if (abs(IAcpx(i))<0.1*Inom)||(abs(IBcpx(i))<0.1*Inom)||
(abs(ICcpx(i))<0.1*Inom)
        SPO(i) = 1;
    else
        SPO(i) = 0;
end

    if (IRB(i)==1)&&(I02(i)==1)&&(SPO(i)==0)
        FIDEN(i) = 1;
    else
FIDEN(i) = 0;
end
end

I0_I2ang = zeros(1,len);
for i = 1:len
```

```
        IO_I2ang(i)  =  angle(IO(i))-angle(I2(i));
        while (IO_I2ang(i)< -pi)
             IO_I2ang(i) = IO_I2ang(i) + 2*pi;
        end
        while (IO_I2ang(i) > pi)
             IO_I2ang(i) = IO_I2ang(i) - 2*pi;
        end
        IO_I2ang(i) = (IO_I2ang(i)/pi)*180;
end
FSA = zeros(1,len);
for i = 1:len
    if ((IO_I2ang(i)>=-30)&&(IO_I2ang(i)<=30))&&(FIDEN(i)==1)
        FSA(i) = 1;
    else
        FSA(i) = 0;
    end
end
FSB = zeros(1,len);
for i = 1:len
    if ((IO_I2ang(i)>=90)&&(IO_I2ang(i)<=150))&&(FIDEN(i)==1)
        FSB(i) = 1;
    else
        FSB(i) = 0;
    end
end
FSC = zeros(1,len);
for i = 1:len
    if ((IO_I2ang(i)>=-150)&&(IO_I2ang(i)<=-90))&&(FIDEN(i)==1)
        FSC(i) = 1;
    else
        FSC(i) = 0;
    end
end
%End
```

The *Relay1_21G_zone1.m* is the pickup logic for Zone 1 Mho ground-distance element. The MATLAB code is shown below. After running this function, if the logic variable Z1G_mho_pu(i) equals 1, it indicates that the Zone 1 Mho ground-distance element picks up.

```
%Relay1_21G_zone1.m, Zone 1 Mho Ground Distance Element
Pickup Logic
%Begin
%Mho AG Element
IA_mag_check = zeros(1,len);
IG_mag_check = zeros(1,len);
MAG_mag_check = zeros(1,len);
MAG1F_pre = zeros(1,len);
MAG1F_pu = zeros(1,len);
for i = 1:len
```

```
    if abs(IAcpx(i)) > Z50G1
        IA_mag_check(i) = 1;
    else
        IA_mag_check(i) = 0;
    end

    if abs(IRcpx(i)) > 0.1*Inom
        IG_mag_check(i) = 1;
    else
        IG_mag_check(i) = 0;
    end

    if (MAG(i)>0)&&(MAG(i) < Z1MG)
        MAG_mag_check(i) = 1;
    else
        MAG_mag_check(i) = 0;
    end

    MAG1F_pre(i) =
IA_mag_check(i)&&IG_mag_check(i)&&MAG_mag_check(i);
    MAG1F_pu(i) = MAG1F_pre(i)&&FSA(i)&&F32Q(i);
end

%Mho BG Element
IB_mag_check = zeros(1,len);
IG_mag_check = zeros(1,len);
MBG_mag_check = zeros(1,len);
MBG1F_pre = zeros(1,len);
MBG1F_pu = zeros(1,len);
for i = 1:len
    if abs(IBcpx(i)) > Z50G1
        IB_mag_check(i) = 1;
    else
        IB_mag_check(i) = 0;
    end

    if abs(IRcpx(i)) > 0.1*Inom
        IG_mag_check(i) = 1;
    else
        IG_mag_check(i) = 0;
    end

    if (MBG(i)>0)&&(MBG(i) < Z1MG)
        MBG_mag_check(i) = 1;
    else
        MBG_mag_check(i) = 0;
    end

MBG1F_pre(i) =
IB_mag_check(i)&&IG_mag_check(i)&&MBG_mag_check(i);
```

```
        MBG1F_pu(i) = MBG1F_pre(i)&&FSB(i)&&F32Q(i);
end

%Mho CG Element
IC_mag_check = zeros(1,len);
IG_mag_check = zeros(1,len);
MCG_mag_check = zeros(1,len);
MCG1F_pre = zeros(1,len);
MCG1F_pu = zeros(1,len);
for i = 1:len
    if abs(ICcpx(i)) > Z50G1
        IC_mag_check(i) = 1;
    else
        IC_mag_check(i) = 0;
    end

    if abs(IRcpx(i)) > 0.1*Inom
        IG_mag_check(i) = 1;
    else
        IG_mag_check(i) = 0;
    end

    if (MCG(i)>0)&&(MCG(i) < Z1MG)
        MCG_mag_check(i) = 1;
    else
        MCG_mag_check(i) = 0;
    end

    MCG1F_pre(i) =
IC_mag_check(i)&&IG_mag_check(i)&&MCG_mag_check(i);
    MCG1F_pu(i) = MCG1F_pre(i)&&FSC(i)&&F32Q;
End
%The overall logic of AG, BG, and CG elements
Z1G_mho_pu = zeros(1,len);
for i=1:len
    Z1G_mho_pu(i) = MAG1F_pu(i)||MBG1F_pu(i)||MCG1F_pu(i);
end
%End
```

Figure 6.14 shows how the pick-up logic values of the AG, BG, and CG elements change during a phase-A-to-ground fault event. Only the AG element picks up the fault, as expected.

The *Relay1_21P_zone1.m* is the pickup logic for Zone 1 Mho phase-distance element. The MATLAB code is shown below. After running this function, if the logic variable $Z1P\_mho\_pu(i)$ equals 1, it indicates that the Zone 1 Mho phase-distance element picks up.

```
%Relay1_21P_zone1.m, Zone 1 Mho Phase Distance Element
Pickup Logic
```

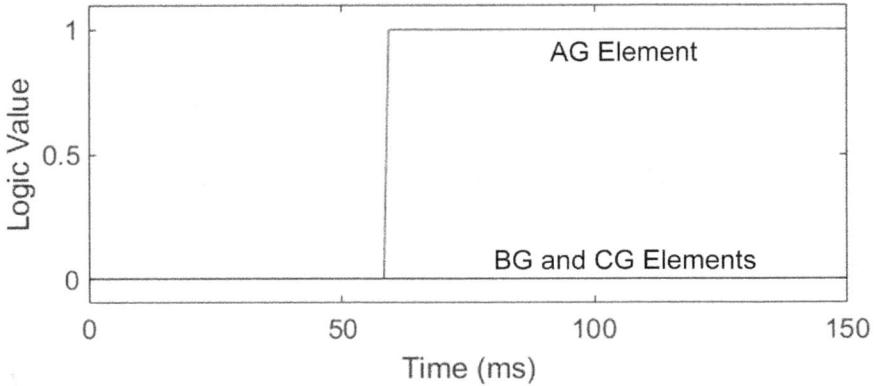

FIGURE 6.14 Pick-up logic values of AG, BG, and CG elements.

```
%Begin
%Mho AB Element
IAB_mag_check = zeros(1,len);
MAB_mag_check = zeros(1,len);
AB_other_check = zeros(1,len);
MAB1F_pre = zeros(1,len);
MAB1F_pu = zeros(1,len);
for i = 1:len
    if abs(IABcpx(i)) > Z50P1
        IAB_mag_check(i) = 1;
    else
        IAB_mag_check(i) = 0;
    end

    if (MAB(i)>0)&&(MAB(i) < Z1MP)
        MAB_mag_check(i) = 1;
    else
        MAB_mag_check(i) = 0;
    end

MAB1F_pre(i) = IAB_mag_check(i)&&MAB_mag_check(i)&&F32Q(i);
    AB_other_check(i) = FSA(i)||FSB(i);
    MAB1F_pu(i) = MAB1F_pre(i) && (~AB_other_check(i));
end

%Mho BC Element
IBC_mag_check = zeros(1,len);
MBC_mag_check = zeros(1,len);
BC_other_check = zeros(1,len);
MBC1F_pre = zeros(1,len);
MBC1F_pu = zeros(1,len);
for i = 1:len
    if abs(IBCcpx(i)) > Z50P1
```

```
             IBC_mag_check(i) = 1;
      else
             IBC_mag_check(i) = 0;
      end

      if  (MBC(i)>0)&&(MBC(i) < Z1MP)
             MBC_mag_check(i) = 1;
      else
             MBC_mag_check(i) = 0;
      end

      MBC1F_pre(i) =
IBC_mag_check(i)&&MBC_mag_check(i)&&F32Q(i);
      BC_other_check(i) = FSB(i)||FSC(i);
      MBC1F_pu(i) = MBC1F_pre(i) && (~BC_other_check(i));
end

%Mho CA Element
ICA_mag_check = zeros(1,len);
MCA_mag_check = zeros(1,len);
CA_other_check = zeros(1,len);
MCA1F_pre = zeros(1,len);
MCA1F_pu = zeros(1,len);
for i = 1:len
      if abs(ICAcpx(i)) > Z50P1
             ICA_mag_check(i) = 1;
      else
             ICA_mag_check(i) = 0;
      end

      if  (MCA(i)>0)&&(MCA(i) < Z1MP)
             MCA_mag_check(i) = 1;
      else
             MCA_mag_check(i) = 0;
      end

      MCA1F_pre(i) =
ICA_mag_check(i)&&MCA_mag_check(i)&&F32Q(i);
      CA_other_check(i) = FSC(i)||FSA(i);
      MCA1F_pu(i) = MCA1F_pre(i) && (~CA_other_check(i));
End
%The overall logic of AB, BC, and CA elements
      Z1P_mho_pu = zeros(1,len);
for i=1:len
Z1P_mho_pu(i) = MAB1F_pu(i)||MBC1F_pu(i)||MCA1F_pu(i);
end
%End
```

The *Relay1_21_zone1.m* is the overall pickup logic for Zone 1 Mho distance element. The MATLAB code is shown below. After running this function, if the

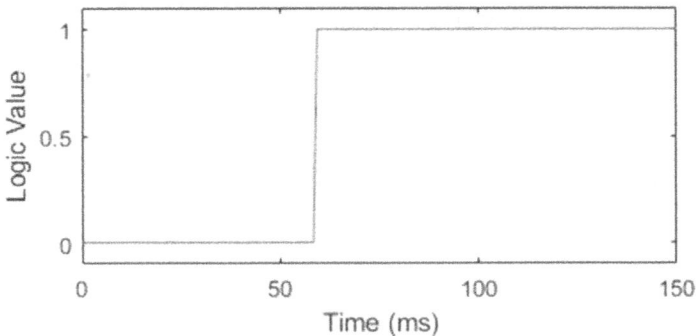

FIGURE 6.15   Value of the Zone 1 pickup logic variable.

logic variable Zone1_mho_pu(i) equals 1, it indicates that the Zone 1 Mho distance element picks up.

```
%Relay1_21_zone1.m, Overall pickup logic for Zone 1 Mho
Distance Element
%Begin
Zone1_mho_pu = zeros(1,len);
for i=1:len
    Zone1_mho_pu(i) = Z1G_mho_pu(i)||Z1P_mho_pu(i);
end
%End
```

Figure 6.15 shows how the logic value of the Zone1_mho_pu variable changes during a phase-A-to-ground fault event.

6.11   SUMMARY

In this chapter, we have mainly illustrated the principles of Mho ground- and phase-distance elements. The concept of voltage polarization and the phenomenon of Mho circle expansion have also been illustrated. The MATLAB programming details of Mho ground- and phase-distance elements have been provided. Readers are encouraged to implement the MATLAB programming code on their own computers to have a better understanding of relay functions and logics.

6.12   PROBLEMS

Problem 6.1

The one-line diagram of a power system is shown in Figure 6.4. The following parameters are given as CT and VT secondary quantities: $E_{SA} = 70\angle 0° \text{ V}$, $E_{SB} = 70\angle 0° \text{ V}$. The positive-sequence impedances: $Z_{SA1} = 1.5\angle 87° \ \Omega$, $Z_{SB1} = 0.8\angle 83° \Omega$, and $Z_{L1} = 5\angle 82° \Omega$. The negative-sequence impedances: $Z_{SA2} = Z_{SA1}$, $Z_{SB2} = Z_{SB1}$, and

$Z_{L2} = Z_{L1}$. The zero-sequence impedances: $Z_{SA0} = 5\angle 87°\,\Omega, Z_{SB0} = 2.5\angle 83°\,\Omega$, and $Z_{L0} = 18\angle 82°\,\Omega$. Current transformer ratio: CTR = 1200/5. Voltage transformer ratio: VTR = 132.8 kV/70 V (line to neutral).

(a) If a phase-A-to-ground fault occurs at the middle point (50%) on the transmission line, calculate the current seen at relay R1 in primary and secondary quantities. You may ignore load flow in this case.

(b) For the conditions in part (a), calculate the effective secondary impedances measured by the AG, AB, and BC elements of relay R1.

Problem 6.2

For the same system described in Problem 6.1: $E_{SB} = 70\angle 30°\,\text{V}$. Consider the load flow.

(a) If a phase-A-to-ground fault occurs at the middle point (50%) on the transmission line, calculate the current seen at relay R1 in primary and secondary quantities. You may use a power system transients simulation program in this problem.

(b) For the conditions in part (a), calculate the effective secondary impedances measured by the AG, AB, and BC elements of relay R1. Compare the results obtained in this problem with the ones obtained in Problem 6.1.

BIBLIOGRAPHY

[1] P. M. Anderson, C. Henville, R. Rifaat, B. Johnson, and S. Meliopoulos, *Power System Protection*, 2nd Ed. Wiley, 2022.

[2] H. J. A. Ferrer and E. O. Schweitzer, editors, *Modern Solutions for Protection, Control, and Monitoring of Electric Power Systems*. Schweitzer Engineering Laboratories, 2010.

[3] H. Lei and B. K. Johnson, "Impact of resistive SFCLs on supervising elements in transmission line protection," *IEEE Transactions on Applied Superconductivity*, vol. 35, no. 5, pp. 1–5, 2025.

[4] E. O. Schweitzer, "New developments in distance relay polarization and fault type selection," in 16th Annual Western Protective Relay Conference, pp. 24–26, October 1989.

[5] D. D. Fentie, "Understanding the dynamic mho distance characteristic," in 69th Annual Conference for Protective Relay Engineers, pp. 1–15, 2016, DOI: 10.1109/CPRE.2016.7914922.

[6] G. E. Alexander and J. G. Andrichak, "Ground distance relaying: problems and principles," in 18th Annual Western Protective Relay Conference, Spokane, WA, October 1991.

[7] E. O. Schweitzer and J. Roberts, "Distance relay element design," in 19th Annual Western Protective Relay Conference, Spokane, WA, October 1992.

[8] R. J. Martilla, "Directional characteristics of distance relay mho elements: part I—a new method of analysis," *IEEE Transactions on Power Apparatus and Systems*, vol. PAS-100, no. 1, pp. 96–102, 1981.

[9] R. J. Marttila, "Directional characteristics of distance relay mho elements: part II-results," *IEEE Transactions on Power Apparatus and Systems*, vol. PAS-100, no. 1, pp. 103–113, 1981.

[10] J. Mooney and J. Peer, "Application guidelines for ground fault protection," in 24th Annual Western Protective Relay Conference, Spokane, WA, October 1997.

[11] A. Iamandi, S. S. Iliescu, N. Arghira, I. Fagarasan, I. Stamatescu, and V. Calofir, "Distance protection scheme for a digital substation," in 2020 IEEE International Conference on Automation, Quality and Testing, Robotics (AQTR), pp. 1–6, 2020, DOI: 10.1109/AQTR49680.2020.9129913.

[12] D. Costello and K. Zimmerman, "Determining the faulted phase," in 63rd Annual Conference for Protective Relay Engineers, pp. 1–20, 2010, DOI: 10.1109/CPRE.2010.5469523.

7 Communication-Aided Protection Schemes

When illustrating distance protection in the previous chapter, we only discussed the relay located at one end of a transmission line. Such single-ended protection schemes have certain limitations in terms of speed and accuracy. In fact, most transmission lines in actual power systems have been equipped with double-ended protection schemes. Communication signals sent from the remote-end relay could accelerate the tripping process of a local relay, especially for a fault occurring near the remote end of the line, or with a large fault resistance. In this chapter, we illustrate four types of commonly used communication-aided protection schemes: direct underreaching transfer trip (DUTT), permissive underreaching transfer trip (PUTT), permissive overreaching transfer trip (POTT), and directional comparison unblocking (DCUB).

7.1 INTRODUCTION

In the previous chapter, we mentioned that a Zone 1 element typically covers 80%–85% of a transmission line. Figure 7.1 shows a case in which a fault occurs at point A, which is beyond the Zone 1 reach and within the Zone 2 reach of the relay at Bus S. If the transmission line is equipped with a single-ended protection scheme only, the relay at Bus S has to wait 20–25 cycles to trip for the fault, which delays the tripping process and could even cause stability issues to the power system. If the relay at Bus R can transmit signal(s) to the relay at Bus S using a communication channel, the tripping process can be accelerated. Such schemes are called communication-aided protection schemes or pilot protection schemes [1]. Signal transmitting via the communication channel could take approximately 1 cycle of time [2]. Depending on specific configurations and trip logics, the transmitted signal(s) could be based on fault detection in Zone 1R or Zone 2R. The transmitted signal could be a direct trip, a permissive trip, a blocking signal, or an unblocking signal. Four types of commonly used communication-aided protection schemes [2] will be illustrated in the following sections.

7.2 DIRECT UNDERREACHING TRANSFER TRIP (DUTT)

The direct underreaching transfer trip (DUTT) logic is shown in Figure 7.2. The subscript "pu" means "pick up". In a DUTT scheme, if a fault is detected within Zone 1 of the relay at Bus R, the circuit breaker at Bus R trips, and a direct trip signal will be sent to the relay on the other end of the line (i.e., Bus S) via a communication channel. It triggers the circuit breaker to trip at Bus S and isolates the

DOI: 10.1201/9781003629481-7

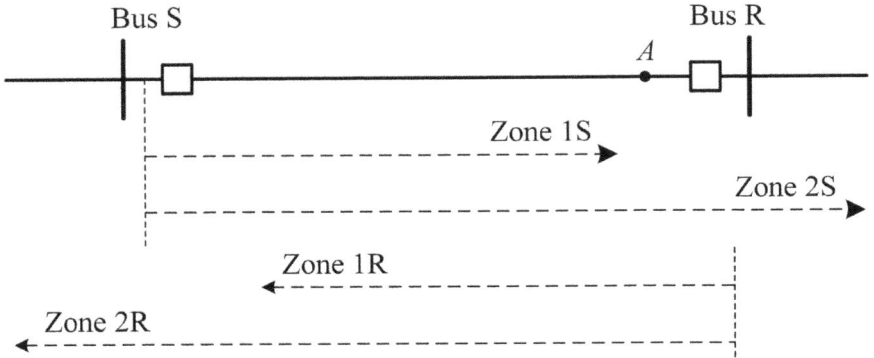

FIGURE 7.1 A two-terminal transmission line.

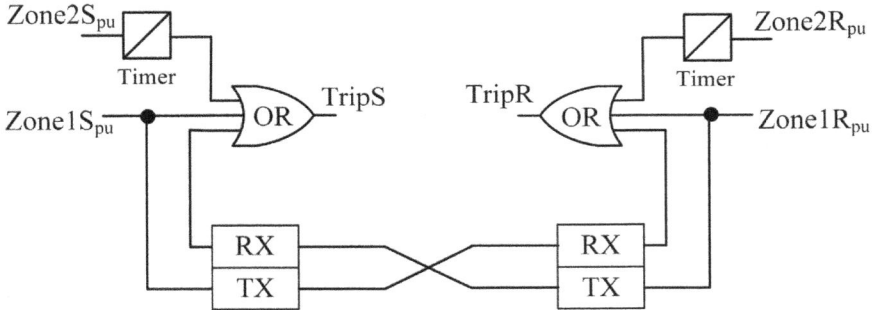

FIGURE 7.2 DUTT logic diagram.

fault quickly. This scheme is simple, but it is susceptible to misoperation because communication channel noise could be misrecognized as a direct trip signal.

7.3 PERMISSIVE UNDERREACHING TRANSFER TRIP (PUTT)

The permissive underreaching transfer trip (PUTT) logic is shown in Figure 7.3. The subscript "pu" means "pick up". In a PUTT scheme, if a fault is detected within Zone 1 of the relay at Bus R, the circuit breaker at Bus R trips, and a permissive trip signal will be sent to the relay on the other end of the line (i.e., Bus S) via a communication channel. The circuit breaker at Bus S trips when it receives the permissive signal and its Zone 2 element detects the fault. This scheme is less susceptible to misoperation than the DUTT scheme because it uses a Zone 2 element to supervise tripping when receiving a permissive signal. The PUTT scheme does not send a permissive signal for out-of-section faults because it uses a Zone 1 element to send a permissive signal.

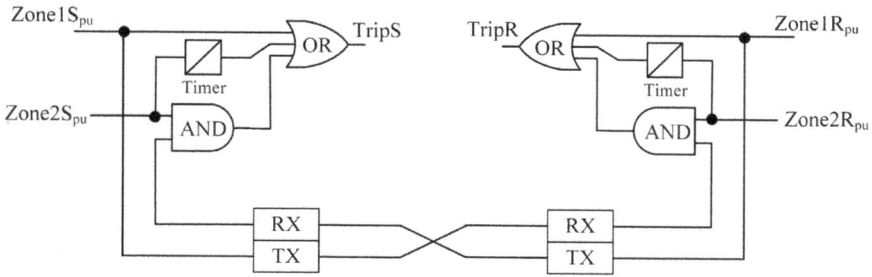

FIGURE 7.3   PUTT logic diagram.

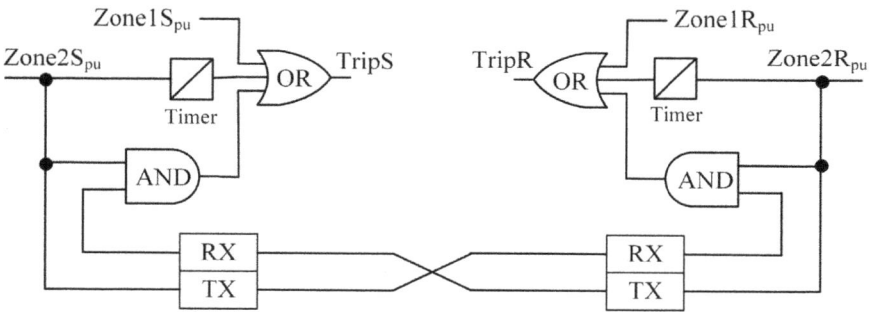

FIGURE 7.4   POTT logic diagram.

7.4   PERMISSIVE OVERREACHING TRANSFER TRIP (POTT)

The permissive overreaching transfer trip (POTT) logic is shown in Figure 7.4. The subscript "pu" means "pick up". Unlike a PUTT scheme, which uses a Zone 1 element to send a permissive signal, a POTT scheme uses a Zone 2 element to send a permissive signal from the local end to the remote end. Because this scheme uses an overreaching (Zone 2) element to send a permissive signal, additional supervisory logic is needed for current reversal situations on parallel lines.

7.5   DIRECTIONAL COMPARISON UNBLOCKING (DCUB)

The DCUB logic is shown in Figure 7.5. The subscript "pu" means "pick up". In a DCUB scheme, the relays located at the two ends of the transmission line continuously send a guard signal to each other. If a fault is detected by the Zone 2 element of a relay, the guard signal will be turned off, and a permissive trip signal will be sent. The relay on the other end of the line will detect the change in signal from guard to permissive trip. In the DCUB scheme, if a loss-of-guard condition is detected and a fault in Zone 2 is also detected by Relay S, it is permitted to trip even if a permissive signal is not received. The response time of a DCUB scheme

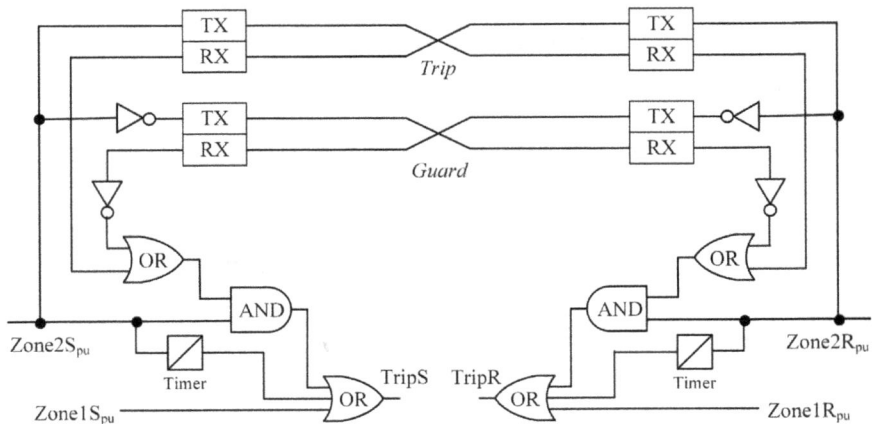

FIGURE 7.5 DCUB logic diagram.

is similar to a POTT scheme because both schemes have a philosophy of tripping with permission from the remote-end Zone 2 element.

7.6 SUMMARY

This chapter introduces four types of commonly used communication-aided protection schemes: direct underreaching transfer trip (DUTT), permissive underreaching transfer trip (PUTT), permissive overreaching transfer trip (POTT), and directional comparison unblocking (DCUB). Their logic diagrams and trip philosophy have been illustrated and compared. One of the main goals of using communication-aided protection schemes is to achieve high-speed tripping for any faults on the transmission line, even with fault resistance. In industry applications, sometimes the features of permissive and blocking schemes can be utilized simultaneously to achieve fast tripping [3].

BIBLIOGRAPHY

[1] P. M. Anderson, C. Henville, R. Rifaat, B. Johnson, and S. Meliopoulos, *Power System Protection*, 2nd Ed. Wiley, 2022.
[2] E. O. Schweitzer III and J. J. Kumm, "Statistical comparison and evaluation of pilot protection schemes," in 23rd Western Protective Relay Conference, pp. 1–20, 1996.
[3] B. Kasztenny, M. V. Mynam, N. Fischer, and A. Guzman, "Permissive or blocking pilot protection schemes? How to have it both ways," in 47th Annual Western Protective Relay Conference, October 2020.

8 Power Swing Blocking and Out-of-Step Tripping

Power swing is a phenomenon when the phase angles of different generators or power sources start to vary relative to each other, which can lead to variations in power flow and instability in the power system. Power swings are often caused by major disturbances such as faults or line switching. Depending on the severity and subsequent control actions, a power swing can be a stable swing in which the system may reach a new stable state, or an unstable swing in which the system may reach an out-of-step condition. An out-of-step condition, also known as loss of synchronism, refers to a condition in which a generator or a portion of the power system loses synchronization with the rest of the system. In this chapter, we illustrate how power swings and out-of-step conditions are typically detected by protective relays.

8.1 INTRODUCTION

The impact of a power swing on relay voltage and current measurements can be illustrated using Figure 8.1. The two sources E_S and E_R represent the equivalent voltage sources of two systems connected to the two ends of the transmission line. The Z_S and Z_R represent the equivalent positive-sequence source impedances. The Z_L represents the positive-sequence line impedance. The V_{1S} and I_1 are the positive-sequence voltage and current measured by the relay at Bus S.

Using the phase angle of E_R as the reference angle, the two voltage sources can be represented as $E_S = |E_S|\angle\theta$ and $E_R = |E_R|\angle 0°$. The current I_1 can be represented using Equation (8.1), in which Z_T is the sum of Z_S, Z_L, and Z_R.

$$I_1 = \frac{|E_S|\angle\theta - |E_R|}{Z_T} \tag{8.1}$$

The positive-sequence impedance Z_1 calculated by the relay at Bus S can be represented using Equation (8.2).

$$Z_1 = \frac{V_{1S}}{I_1} = \frac{|E_S|\angle\theta - I_1 Z_S}{I_1} = \frac{|E_S|\angle\theta}{|E_S|\angle\theta - |E_R|} Z_T - Z_S \tag{8.2}$$

If $|E_S| = |E_R|$, Equation (8.2) can be simplified as Equation (8.3).

$$Z_1 = \frac{Z_T}{2}\left(1 - jcot\frac{\theta}{2}\right) - Z_S \tag{8.3}$$

DOI: 10.1201/9781003629481-8

95

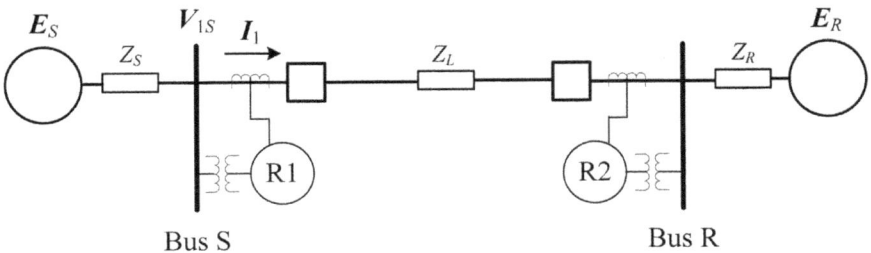

FIGURE 8.1   A power system for power swing illustration.

If the two sources have no slip frequency, the angle θ stays unchanged. Under normal operating conditions, angle θ is not large (typically less than 45°), and the positive-sequence impedance Z_1 calculated by the relay stays outside Mho circles. When a power swing occurs, the angle θ could change, and the positive-sequence impedance Z_1 calculated by the relay could enter Mho circles, which could potentially cause the relay's distance elements misoperate. In fact, a power swing not only increases misoperation risks to distance elements, but the voltage and current fluctuations during a power swing could cause misoperations of phase overcurrent, directional overcurrent, phase overvoltage, and undervoltage elements [1].

Example 8.1

The one-line diagram of a simple power system is shown in Figure 8.1. The system's primary nominal voltage, $V_{LLprimary} = 230\,\text{kV}$. The nominal frequency, $f_{nom} = 60\,\text{Hz}$. The nominal angular frequency, $\omega_s = 2\pi f_{nom}$. The voltage transformer ratio, VTR = 2000. The current transformer ratio, CTR = 160. The primary positive-sequence line impedance, $Z_{Lpri} = 100$ Ohm with 87.6 degrees. The sending source's positive-sequence impedance, $Z_S = 0.25 Z_{Lpri}$. The receiving source's positive-sequence impedance $Z_R = 0.25 Z_{Lpri}$. The voltage magnitude E_S is 3% above nominal. The voltage magnitude E_R is 3% below nominal. The angle between E_S and E_R is $\delta_{pre} = 40$ degrees.

(a) Calculate the nominal secondary voltage of the system.
(b) Calculate E_S secondary voltage.
(c) Calculate E_R secondary voltage.
(d) Calculate secondary line impedance.
(e) Calculate the voltage and current at the relay R1 location for the 40-degree load angle.
(f) Calculate the apparent impedance seen by the relay R1 for the 40-degree load angle.
(g) Calculate the three-phase apparent power seen by the relay R1.

Solution:

(a) The nominal secondary voltage (line to neutral) of the system:

$$V_{LGsec} = \frac{V_{LLprmary}}{VTR\sqrt{3}} = 66.395 \ V$$

(b) The E_S secondary voltage:

$$E_{Ssec} = 1.03 * 66.395 * e^{j40 \ degrees} = 68.387\angle40° \ V$$

(c) The E_R secondary voltage:

$$E_{Rsec} = 0.97 * 66.395 * e^{j0 \ degree} = 64.403\angle0° \ V$$

(d) The secondary line impedance:

$$Z_{Lsec} = Z_{Lpri}\frac{CTR}{VTR} = 8\angle87.6° \ \Omega$$

The secondary source impedances:

$$Z_{Ssec} = 0.25Z_{Lsec} = 2\angle87.6° \ \Omega$$

$$Z_{Rsec} = 0.25Z_{Lsec} = 2\angle87.6° \ \Omega$$

(e) $Z_T = Z_{Ssec} + Z_{Rsec} + Z_{Lsec} = 12\angle87.6° \ \Omega$

The secondary current at the relay R1 location:

$$I_{1sec} = \frac{E_{Ssec} - E_{Rsec}}{Z_T} = 3.798\angle17.688° \ A$$

The secondary voltage at the relay R1 location:

$$V_{1sec} = E_{Ssec} - I_{1sec}Z_{Ssec} = 65.576\angle33.96° \ V$$

(f) The apparent impedance seen by the relay R1:

$$Z_{appsec} = \frac{V_{1sec}}{I_{1sec}} = 17.268\angle16.272° \ \Omega$$

Converting to the primary side,

$$Z_{apppri} = Z_{appsec}\frac{VTR}{CTR} = 215.85\angle16.272° \ \Omega$$

(g) The primary current at the relay R1 location:

$$I_{1pri} = I_{1sec} * CTR = 607.68\angle17.688° \ A$$

The primary voltage at the relay R1 location:

$$V_{1pri} = V_{1sec} * VTR = 131.152\angle33.96° \ kV \ line \ to \ neutral.$$

The three-phase apparent power:

$$S_{3ph} = 3|V_{1pri}| \cdot |I_{1pri}| = 239.095 \ MVA$$

8.2 STABLE AND UNSTABLE POWER SWINGS

Figure 8.2 shows an example of stable and unstable power swings. A stable power swing could return to a new stable state, and an unstable power swing could cause out-of-step conditions. Both stable and unstable power swings could enter the distance element operating characteristics. Zone 1 distance elements are particularly

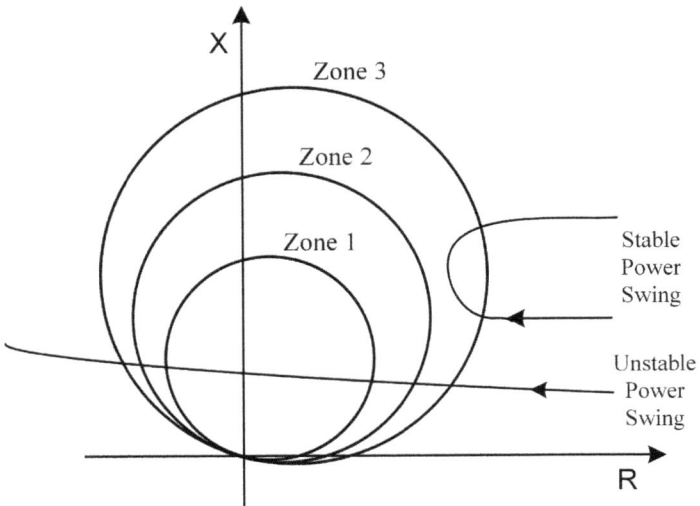

FIGURE 8.2   Stable and unstable power swings.

sensitive to power swings. The measured impedance typically changes slowly during power swings. A Zone 2 or Zone 3 element could misoperate if the measured impedance stays inside a Zone 2 or Zone 3 Mho circle for a time longer than the time delay setting.

The general protection philosophy for treating power swings is to avoid tripping for stable power swings and to initiate system islanding for unstable power swings to prevent further blackouts. For either stable or unstable power swings, relay elements prone to misoperate (e.g., distance elements) should be blocked. The power swing blocking (PSB) function differentiates between faults and power swings, and blocks certain relay elements if a power swing is confirmed. The ANSI/IEEE number for power swing blocking is 68. The out-of-step tripping (OOST) function differentiates between unstable and stable power swings and issues tripping signal(s) if an unstable power swing is confirmed. The ANSI/IEEE number for out-of-step tripping is 78.

8.3   POWER SWING DETECTION

The rate of change of the measured impedance can be utilized to detect power swings [2]. The principle is illustrated using Figure 8.3. The characteristic consists of two concentric polygons, one of which is called outer power swing detection (OPSD) blinder and the other is called inner power swing detection (IPSD) blinder. The IPSD blinder is set to be outside the largest Mho circle characteristic that we plan to block. Also, the IPSD blinder should be set to avoid tripping for the worst stable power swing if possible. The OPSD blinder should be set to avoid tripping for the maximum load conditions.

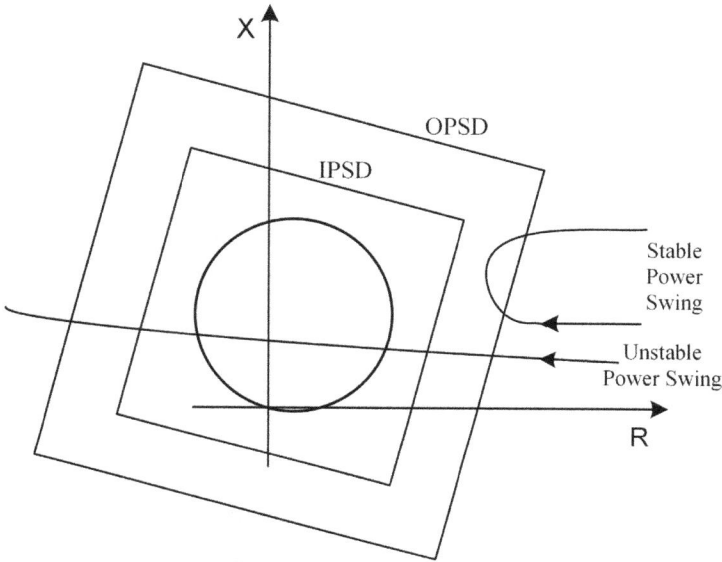

FIGURE 8.3 Power swing detection using two concentric polygons.

The measured impedance changes very fast during faults and changes slowly during power swings. Based on the measured impedance rate of change, the power swing blocking (PSB) function could differentiate between faults and power swings. A blocking signal will be issued by the PSB function if a measured positive-sequence impedance stays inside the area between the OPSD and IPSD blinders for a time longer than a blocking delay setting T_{block}.

The measured impedance changes faster during unstable power swings than during stable power swings, which can be used to differentiate between unstable power swings and stable power swings. An unstable power swing is confirmed if a measured positive-sequence impedance stays between the OPSD and IPSD blinders longer than a tripping delay setting T_{trip} and shorter than T_{block}, and then enters the IPSD boundary. Then, the OOST function will issue a tripping signal to initiate system islanding actions. The out-of-step tripping (OOST) function can issue a tripping signal when the measured positive-sequence impedance enters the IPSD boundary. This scheme is called trip-on-the-way-in (TOWI). The out-of-step tripping (OOST) function can also issue a tripping signal when the measured positive-sequence impedance exits the IPSD and OPSD blinders. This scheme is called trip-on-the-way-out (TOWO). TOWO generally causes less stress than TOWI on circuit breakers.

Besides the method illustrated earlier, other methods are also available for power swing detection, such as continuous impedance calculation, continuous incremental current calculation, and swing-center-voltage (SCV) methods. Readers are encouraged to review pertinent references [3–7] for more details.

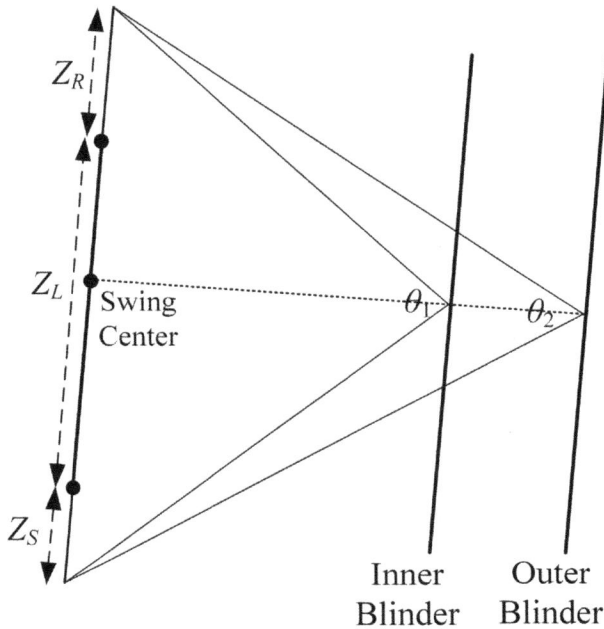

FIGURE 8.4   Swing center and two blinders.

Example 8.2

The one-line diagram of a simple power system is shown in Figure 8.1. The nominal frequency, $f_{nom} = 60$ Hz. The nominal angular frequency, $\omega_s = 2\pi f_{nom}$. The positive-sequence line impedance $Z_L = 8$ Ohm with 87.7 degrees. The sending source's positive-sequence impedance $Z_S = 1.776$ Ohm with 87.7 degrees. The receiving source's positive-sequence impedance $Z_R = 3.3$ Ohm with 87.7 degrees. All the impedances are given as secondary quantities. The secondary voltage magnitude, $E_S = 61.4$ V. The secondary voltage magnitude E_R is 80% of E_S. The initial angle between E_S and E_R is $\delta_{pre} = 50$ degrees. As shown in Figure 8.4, the inner blinder load angle, $\theta_1 = 90$ degrees. The outer blinder load angle, $\theta_2 = 60$ degrees. There is a frequency difference between E_S and E_R, which is also called slip frequency. The slip frequency, $f_{slip} = 3$ Hz.

(a) Determine the apparent impedance seen by the relay R1 when the angle between E_S and E_R is $50°$.
(b) Determine the time duration that the apparent impedance stays between the outer and inner blinders. This time duration is also called the out-of-step blocking delay (OSBD).
(c) The Zone 2 reach of relay R1 is 120% of the line impedance. Plot the swing locus of the apparent impedance seen by the relay R1. Also, plot the two blinders and the Zone 2 Mho circle.

Solution:

(a) $E_S = 61.4\angle 50°$ V

$E_R = 0.8|E_S|\angle 0° = 49.15\angle 0°$ V

$Z_T = Z_S + Z_L + Z_R = 13.07\angle 87.7°$ Ω

The secondary current at the relay R1 location:

$$I_1 = \frac{E_S - E_R}{Z_T} = 3.67\angle 13.84°\ A$$

The secondary voltage at the relay R1 location:

$V_1 = E_S - I_1 Z_S = 57.6\angle 44.89°$ V

The apparent impedance seen by the relay R1:

$$Z_{app} = \frac{V_1}{I_1} = 15.68\angle 31.05°\ \Omega$$

(b) The out-of-step blocking delay (OSBD):

$$OSBD = \frac{(\theta_1 - \theta_2)}{(360\deg) f_{slip}} = \frac{30}{360*3} = 2.78\ ms$$

(c) The distance between the inner blinder and the swing center:

$$d_1 = \frac{|Z_T|}{2\tan(\theta_1 / 2)} = 6.54\ \Omega$$

The distance between the outer blinder and the swing center:

$$d_2 = \frac{|Z_T|}{2\tan(\theta_2 / 2)} = 11.32\ \Omega$$

Zone 2 impedance, $Z2P = 1.2|Z_L| = 9.6\ \Omega$.

The swing locus, two blinders, line impedance, and Zone 2 Mho circle are plotted in Figure 8.5 using Mathcad. The real and imaginary axes are also plotted in Figure 8.5. From Figure 8.5, we can see that the swing locus passes through the Zone 2 Mho circle.

8.4 SUMMARY

This chapter briefly illustrates how power swings and out-of-step conditions are typically detected by protective relays. Based on the rate of change of the measured impedance, protective relays can differentiate between faults and power swings, and between stable power swings and unstable power swings.

8.5 PROBLEMS

Problem 8.1

Continue from Example 8.1, set the Zone 2 reach of relay R1 as 120% of the line impedance. Plot the line impedance and the Zone 2 Mho circle in the impedance plane.

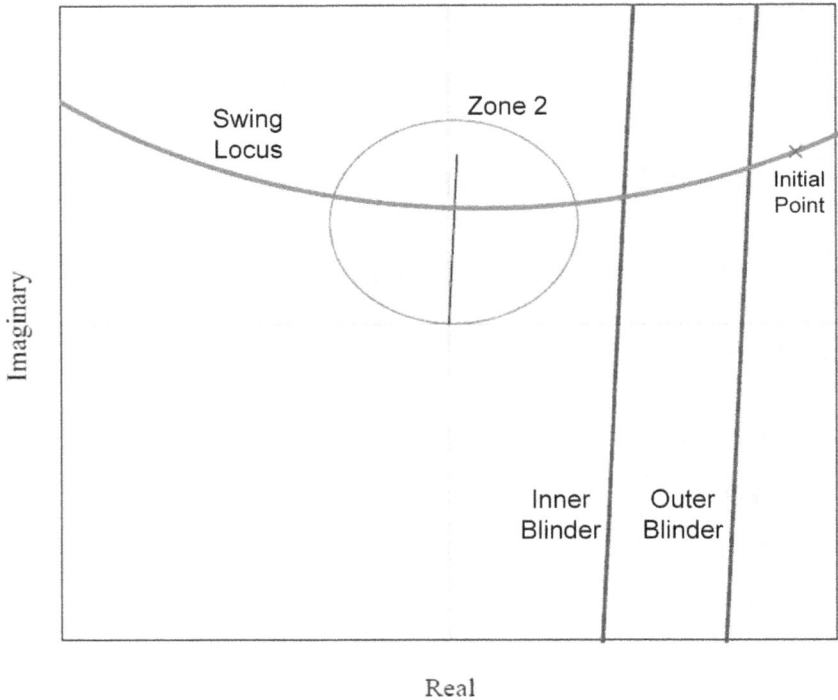

FIGURE 8.5 Plots of swing locus and Zone 2 mho circle.

Problem 8.2

Continue from Example 8.1, there is a frequency difference between E_S and E_R, which is also called slip frequency. If the slip frequency $f_{slip} = 4$ Hz, plot the swing locus of the apparent impedance seen by the relay R1.

BIBLIOGRAPHY

[1] H. J. A. Ferrer and E. O. Schweitzer, editors, *Modern Solutions for Protection, Control, and Monitoring of Electric Power Systems.* Schweitzer Engineering Laboratories, 2010.

[2] D. A. Tziouvaras and D. Hou, "Out-of-step protection fundamentals and advancements," in 57th Annual Conference for Protective Relay Engineers, pp. 282–307, 2004, DOI: 10.1109/CPRE.2004.238495.

[3] K. Sreenivasachar, "Out-of-step detection on transmission lines using apparent impedance differential method," *IEEE Transactions on Power Delivery*, vol. 37, no. 4, pp. 3245–3256, 2021.

[4] V. A. Ambekar and S. S. Dambhare, "Comparative evaluation of out of step detection schemes for distance relays," in 2012 IEEE 5th Power India Conference, pp. 1–6, 2012.

[5] M. Afzali and A. Esmaeilian, "A novel algorithm to identify power swing based on superimposed measurements," in 2012 11th International Conference on Environment and Electrical Engineering, pp. 1109–1113, 2012.

[6] J. G. Rao and A. K. Pradhan, "Power-swing detection using moving window averaging of current signals," *IEEE Transactions on Power Delivery*, vol. 30, no. 1, pp. 368–376, 2014.

[7] J. Mooney and N. Fischer, "Application guidelines for power swing detection on transmission systems," in 2006 Power Systems Conference: Advanced Metering, Protection, Control, Communication, and Distributed Resources, pp. 159–168, 2006.

9 Differential Protection

Differential protection is an effective scheme for detecting faults inside the protection zone of a substation bus, a transformer, or a transmission line. The general process of differential protection is comparing the current flowing into the protected zone with the current flowing out. An internal fault is detected if there is a significant unbalance between the two compared currents. Inherently, differential protection is based on the principle that under normal conditions, the total outgoing current should be approximately equal to the total incoming current according to Kirchhoff's current law. In a transformer case, the current ratio should be taken into account when applying this principle. In this chapter, we illustrate the principles of differential protection and practical considerations in applications.

9.1 INTRODUCTION

A diagram including multiple differential protection zones is shown in Figure 9.1. These protection zones typically have overlaps [1]. Such overlaps ensure that even equipment located near the boundary of a zone is protected, preventing any gaps. The overlapping zones provide redundancy in the protection system, increasing the overall reliability and security of the protection system [2–4]. If the primary protection within a zone fails to operate correctly, the overlapping zone provides a backup protection to isolate the fault.

The general principle of differential protection can be illustrated using Figure 9.2. The I_{1ABC}, I_{2ABC}, and I_{3ABC} represent the three-phase currents on branches 1, 2, and 3 connected to the equipment protected, with the direction of coming into the equipment defined as the positive direction. Under normal load conditions and external fault conditions (e.g., a fault occurring at point F), the net incoming current is approximately zero in each phase, as shown in Equation (9.1).

$$\begin{cases} I_{1A} + I_{2A} + I_{3A} \approx 0 \\ I_{1B} + I_{2B} + I_{3B} \approx 0 \\ I_{1C} + I_{2C} + I_{3C} \approx 0 \end{cases} \tag{9.1}$$

If an internal fault (e.g., a phase-A-to-ground fault) occurs, the net incoming current in the faulted phase(s) will be significantly different from zero, as shown in Equation (9.2).

$$I_{1A} + I_{2A} + I_{3A} = I_{Af} \tag{9.2}$$

Based on the current features of internal fault conditions, we can set a threshold to differentiate between an internal fault and an external fault.

DOI: 10.1201/9781003629481-9

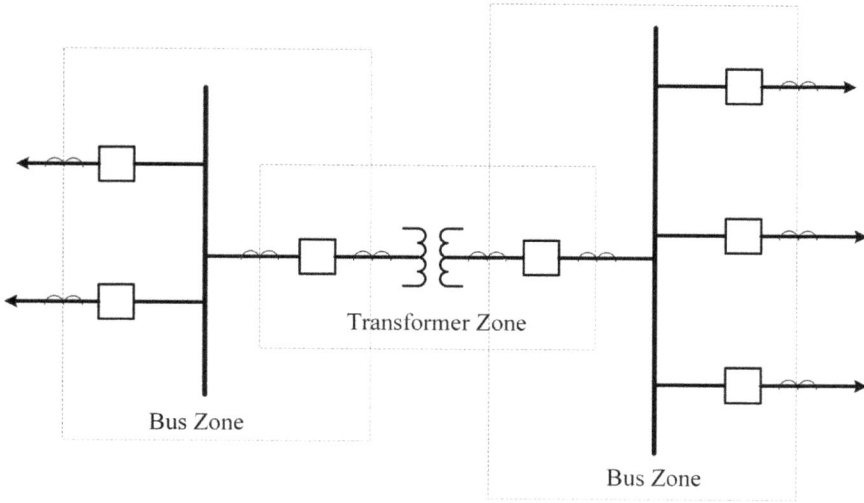

FIGURE 9.1 Multiple protection zones in a substation.

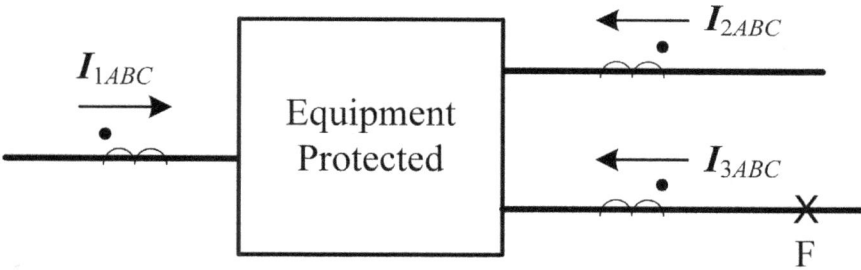

FIGURE 9.2 General principle of differential protection.

9.2 BUS DIFFERENTIAL PROTECTION

The general principle of differential protection has been illustrated in Section 9.1. Still, some practical considerations need to be taken into account in applications [5]. It should be noted that relays operate on CT secondary currents. Thus, in a bus differential protection scheme, the CTs installed on all branches should have matched parameters.

CT saturation brings about concerns, especially for external faults. Ideally, the net incoming current is approximately zero in each phase under external faults. However, the net incoming current at CT secondary may be significantly different from zero due to CT saturation, as illustrated in Example 9.1. This brings challenges to differentiating between an external fault and an internal fault.

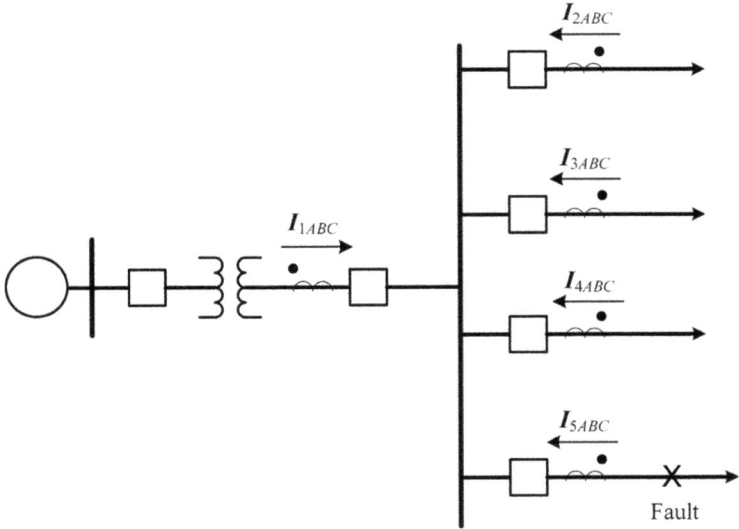

FIGURE 9.3 One-line diagram for Example 9.1.

Example 9.1

The one-line diagram of a substation bus with five branches is shown in Figure 9.3. An external phase-A-to-ground fault occurs on branch 5. The CT installed on phase A of branch 5 has deep saturation due to high current. Because of the CT saturation, the measured CT secondary current amplitude reduces to 50% of its actual amplitude.

On the primary side, the relationship $I_{1A} + I_{2A} + I_{3A} + I_{4A} + I_{5A} \approx 0$ is still valid. On the CT secondary side, the relay sees 50% of the actual current on branch 5. It will calculate $I_{1A} + I_{2A} + I_{3A} + I_{4A} + 0.5I_{5A}$, which is significantly different from zero.

Several options are available to deal with CT saturation. One option is to use linear CTs with no iron cores to avoid CT saturation. Linear CTs include linear couplers and optical CTs. Optical CTs are based on the Faraday effect, which is a phenomenon where the polarization plane of light rotates when it propagates through a material under the influence of a magnetic field. Optical CTs utilize this effect by measuring the rotation of light polarization caused by the magnetic field generated by the current flowing through a conductor [6]. Another option is to use a low-impedance differential or a high-impedance differential protection scheme [7].

9.2.1 LOW-IMPEDANCE DIFFERENTIAL PROTECTION

Low-impedance differential is also called restrained differential or percentage differential. Instead of directly calculating the sum of all incoming currents and comparing the net incoming current with zero, a low-impedance differential protection scheme calculates two current values—the operating current I_{OP} and the restraint current I_{RT}.

The operating and restraint currents are calculated for each individual phase. Using the phase A as an example, the operating and restraint currents are calculated using Equations (9.3) and (9.4), respectively. The subscript "A" means phase A, and the subscript "n" means the total number of branches connected to the bus.

$$I_{OP} = \left| I_{1A} + I_{2A} + I_{3A} + \ldots + I_{nA} \right| \tag{9.3}$$

$$I_{RT} = \left| I_{1A} \right| + \left| I_{2A} \right| + \left| I_{3A} \right| + \ldots + \left| I_{nA} \right| \tag{9.4}$$

The ratio of I_{OP} to I_{RT} is compared with a threshold K_0. An internal fault is confirmed if the ratio is greater than K_0, as shown in Equation (9.5).

$$\frac{I_{OP}}{I_{RT}} > K_0 \tag{9.5}$$

The ratios of I_{OP} to I_{RT} under load, external fault, and internal fault conditions are plotted in Figure 9.4.

It is expected that the severity of CT saturation varies with different current levels. CT saturation typically becomes more severe with a higher current level. A dual slope characteristic curve or dynamic slope setting can be used in applications, as shown in Figure 9.5 (a) and (b), respectively.

For the dual-slope characteristic, a smaller slope threshold is used for lower current levels, and a larger slope threshold is used for higher current levels. For

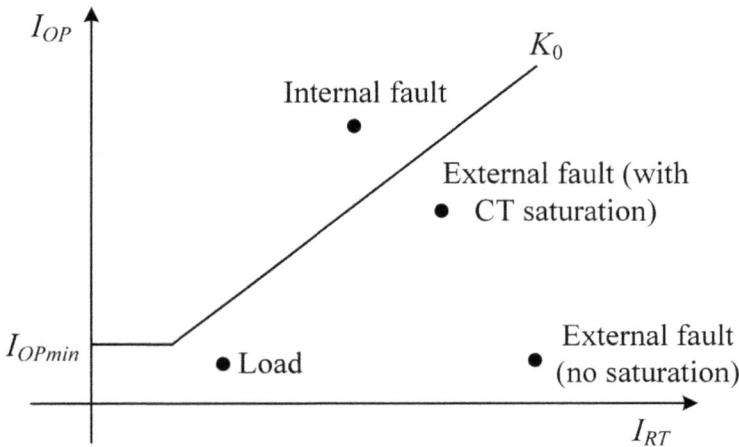

FIGURE 9.4   Restrained differential protection characteristic curve.

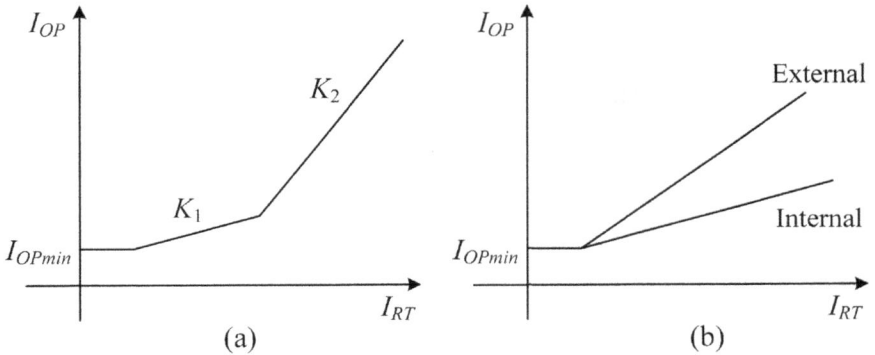

FIGURE 9.5 (a) Dual-slope characteristic and (b) dynamic slope setting.

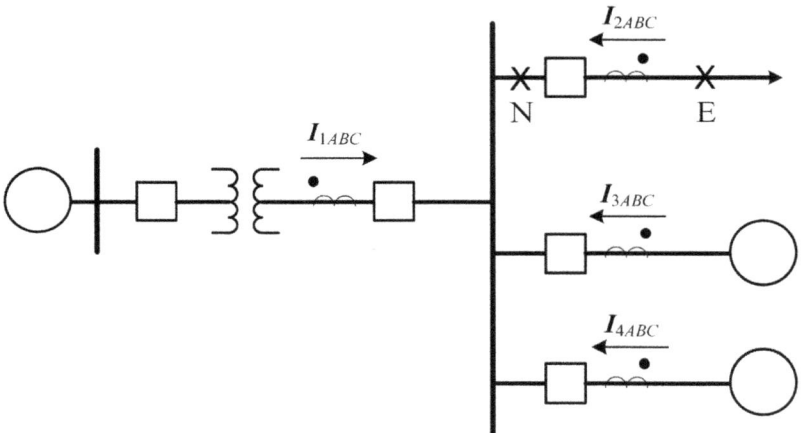

FIGURE 9.6 One-line diagram for Example 9.2.

the dynamic slope setting, a smaller slope threshold is used for internal fault detection, which is more sensitive. A larger slope threshold is used for external fault restraint, which is more secure.

Example 9.2

The one-line diagram of a substation bus with four branches is shown in Figure 9.6.

(a) An external phase-A-to-ground fault occurs at point E on branch 2. No CT saturation occurs. The phase A secondary currents measured on branches 1, 2, 3, and 4 are $40 \angle -90°$ A, $150 \angle 90°$ A, $50 \angle -90°$ A, and $60 \angle -90°$ A, respectively. Calculate the I_{OP} and I_{RT}.

(b) An external phase-A-to-ground fault occurs at point E on branch 2. The CT installed on phase A of branch 2 has saturation. Because of CT saturation, the measured CT secondary current amplitude reduces to 2/3 of the value measured in part (a), assuming there is no phase shift. The phase A secondary currents measured on other branches stay the same as part (a). Calculate the I_{OP} and I_{RT}.

(c) An internal phase-A-to-ground fault occurs at point N on branch 2. The phase A secondary currents measured on branches 1, 2, 3, and 4 are $40\angle -90°$ A, 0 A, $50\angle -90°$ A, and $60\angle -90°$ A, respectively. Calculate the I_{OP} and I_{RT}.

Solution:

(a) External fault with no CT saturation:

$$I_{OP} = \left|40\angle -90° + 150\angle 90° + 50\angle -90° + 60\angle -90°\right| = 0\,\text{A}$$

$$I_{RT} = \left|40\angle -90°\right| + \left|150\angle 90°\right| + \left|50\angle -90°\right| + \left|60\angle -90°\right| = 300\,\text{A}$$

(b) External fault with CT saturation:

$$I_{OP} = \left|40\angle -90° + 100\angle 90° + 50\angle -90° + 60\angle -90°\right| = 50\,\text{A}$$

$$I_{RT} = \left|40\angle -90°\right| + \left|100\angle 90°\right| + \left|50\angle -90°\right| + \left|60\angle -90°\right| = 250\,\text{A}$$

(c) Internal fault:

$$I_{OP} = \left|40\angle -90° + 0 + 50\angle -90° + 60\angle -90°\right| = 150\,\text{A}$$

$$I_{RT} = \left|40\angle -90°\right| + \left|50\angle -90°\right| + \left|60\angle -90°\right| = 150\,\text{A}$$

The ratios of I_{OP}/I_{RT} in cases (a), (b), and (c) are 0, 0.2, and 1.0, respectively. If we set the slope $20\% < K_0 < 80\%$, it could differentiate between an external fault and an internal fault.

9.2.2 High-Impedance Differential Protection

The principle of high-impedance differential protection is very different from restrained differential (low-impedance differential) protection. The general principle of high-impedance differential protection is to expect a CT to saturate for external faults. If a CT starts to saturate during an external fault, it will be driven into deep saturation. This principle can be illustrated using Figure 9.7.

Figure 9.7 (a) shows an external fault example. An external fault occurs at point E on branch 2. The current on branch 2 CT approximately equals the total current from other CTs. The branch 2 CT starts to saturate. Once the CT starts to saturate, all currents from other branches will be driven into this saturation branch, causing the CT to saturate deeper. As a result, the equivalent impedance of branch 2 CT is reduced. Also, there is not enough current flowing into branch 5. The magnitude of the voltage across the branch 5 resistor is below the relay tripping threshold. The relay will not trip.

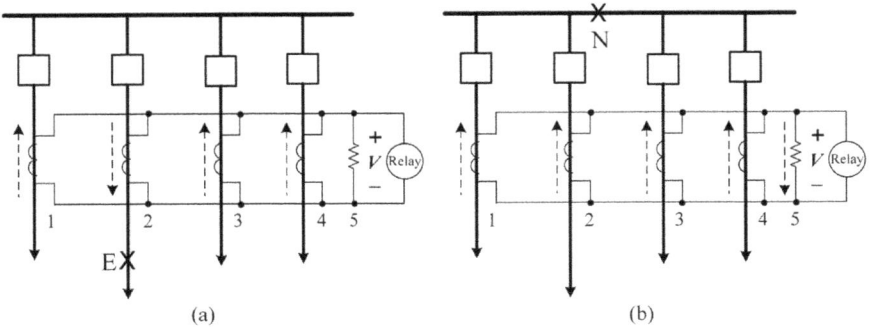

FIGURE 9.7   Principle of high-impedance differential protection.

Figure 9.7 (b) shows an internal fault example. An internal fault occurs at point N on the bus. The current on branch 5 approximately equals the total current from all branches with CTs. The magnitude of the voltage across the branch 5 resistor will exceed the relay tripping threshold. The relay will trip.

It should be noted that the parameters of CTs are carefully selected so that deep saturation occurs only for external faults and little or no saturation occurs for internal faults. Before selecting CT parameters, thorough fault studies for the system need to be performed.

9.3   TRANSFORMER DIFFERENTIAL PROTECTION

Transformers are vital and expensive components in electric power systems. Failure of transformers could cause long-time consequences. In this section, we will mainly illustrate the principle of transformer differential protection, which is effective in protecting transformers from internal faults. Differential protection is the first line for transformer protection. Besides internal faults, other types of transformer stresses and faults include, but are not limited to, overexcitation, winding overcurrent, heating due to overload or inadequate cooling, and through faults. Readers are encouraged to check IEEE C37.91—IEEE guide for protecting power transformers [8], for more details.

The principle of transformer differential protection is similar to bus differential protection, with some differences. The turn ratio of a transformer is affected by specific taps connected. Delta-to-Wye and Wye-to-Delta connections will cause phase shift and also affect the turn ratio. Besides, for a Wye-grounded-to-Delta transformer, the Wye side has zero-sequence current during ground faults, whereas the Delta side does not. These factors must be taken into account in transformer differential protection design and applications. For electromechanical relays, the CTs are connected properly to cancel the phase shift from the transformer primary to secondary windings. For digital relays (i.e., microprocessor-based relays), algorithms with matrix multiplications are performed to compensate for the phase shift and remove the zero-sequence current. Detailed

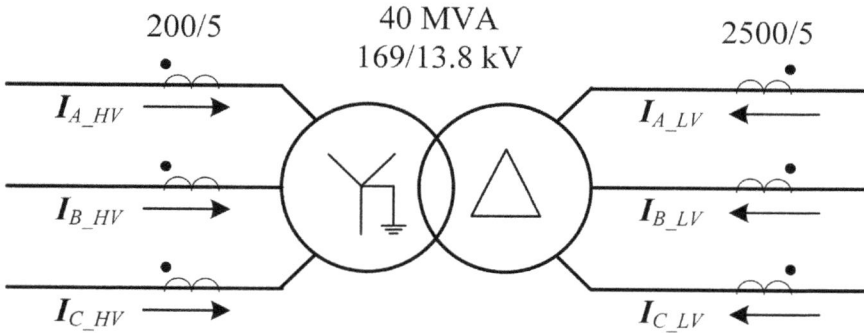

FIGURE 9.8   Diagram for Example 9.3.

procedures are shown in Example 9.3 to illustrate how the matrix multiplications are performed.

Example 9.3

A 169/13.8 kV, 40 MVA, Yd1 transformer is shown in Figure 9.8. The Y side is grounded. The transformer high-voltage (HV) side line currents are $I_{A\_HV} = 100\angle 0°$ A, $I_{B\_HV} = 100\angle -120°$ A,   and $I_{C\_HV} = 100\angle 120°$ A.   The transformer low-voltage (LV) side line currents are $I_{A\_LV} = 1224.6\angle 150°$ A, $I_{B\_LV} = 1224.6\angle 30°$ A , and $I_{C\_LV} = 1224.6\angle -90°$ A. All the currents are given as CT primary. The direction of currents is shown in the figure. The transformer high-voltage-side CT ratio, $CTR_{HV} = 200/5$. The transformer low-voltage-side CT ratio, $CTR_{LV} = 2500/5$. Both sets of CTs are connected in Wye. The transformer is protected by a digital relay.

During routine maintenance, a technician accidentally shorts out the phase C CT on the HV side. Determine whether or not the transformer differential element(s) will operate. If so, which phase differential element(s) will operate?

Solution:

$$S_{rated} = 40 \text{ MVA}, V_{HL} = 169 \text{ kV}, V_{LV} = 13.8 \text{ kV}.$$

We will calculate TAP currents so that, under full load conditions, 1.0 per unit phase current is measured by the relay. The TAP currents are essentially CT secondary base currents for per-unit calculation.

The transformer HV-side TAP current, $TAP_{HV} = \dfrac{S_{rated}}{\sqrt{3}V_{HV}CTR_{HV}} = 3.416$ A.

The transformer LV-side TAP current, $TAP_{LV} = \dfrac{S_{rated}}{\sqrt{3}V_{LV}CTR_{LV}} = 3.347$ A.

The following compensation matrices MAT_{11} and MAT_0 will be used to correct the phase shift caused by the Wye–Delta transformer.

$$MAT_{11} = \frac{1}{\sqrt{3}} \begin{bmatrix} 1 & 0 & -1 \\ -1 & 1 & 0 \\ 0 & -1 & 1 \end{bmatrix} \qquad\qquad MAT_0 = \begin{bmatrix} 1 & 0 & 0 \\ 0 & 1 & 0 \\ 0 & 0 & 1 \end{bmatrix}$$

The uncorrected CT secondary currents:

$$\begin{bmatrix} I_{A\_HV\_sec} \\ I_{B\_HV\_sec} \\ I_{C\_HV\_sec} \end{bmatrix} = \frac{1}{CTR_{HV}} \begin{bmatrix} I_{A\_HV} \\ I_{B\_HV} \\ I_{C\_HV} \end{bmatrix} \qquad \begin{bmatrix} I_{A\_LV\_sec} \\ I_{B\_LV\_sec} \\ I_{C\_LV\_sec} \end{bmatrix} = \frac{1}{CTR_{LV}} \begin{bmatrix} I_{A\_LV} \\ I_{B\_LV} \\ I_{C\_LV} \end{bmatrix}$$

The corrected CT secondary currents:

$$\begin{bmatrix} I_{A\_HV\_cor} \\ I_{B\_HV\_cor} \\ I_{C\_HV\_cor} \end{bmatrix} = \frac{1}{TAP_{HV}} MAT_{11} \begin{bmatrix} I_{A\_HV\_sec} \\ I_{B\_HV\_sec} \\ I_{C\_HV\_sec} \end{bmatrix} \qquad \begin{bmatrix} I_{A\_LV\_cor} \\ I_{B\_LV\_cor} \\ I_{C\_LV\_cor} \end{bmatrix} = \frac{1}{TAP_{LV}} MAT_0 \begin{bmatrix} I_{A\_LV\_sec} \\ I_{B\_LV\_sec} \\ I_{C\_LV\_sec} \end{bmatrix}$$

The corrected CT secondary currents are per-unit currents. It should be noted that there is not only one option of matrices for transformer phase shift correction but using other matrices may also achieve the same objective.

The operating and restraint currents:

Phase A:

$$I_{OPA} = \left| I_{A\_HV\_cor} + I_{A\_LV\_cor} \right| = 2.252 \times 10^{-5}$$

$$I_{RTA} = \left| I_{A\_HV\_cor} \right| + \left| I_{A\_LV\_cor} \right| = 1.464$$

Phase B:

$$I_{OPB} = \left| I_{B\_HV\_cor} + I_{B\_LV\_cor} \right| = 2.252 \times 10^{-5}$$

$$I_{RTB} = \left| I_{B\_HV\_cor} \right| + \left| E_{B\_LV\_cor} \right| = 1.464$$

Phase C:

$$I_{OPC} = \left| I_{C\_HV\_cor} + I_{C\_LV\_cor} \right| = 2.252 \times 10^{-5}$$

$$I_{RTC} = \left| I_{C\_HV\_cor} \right| + \left| I_{C\_LV\_cor} \right| = 1.464$$

If the phase-C CT on the HV side is shorted:

The uncorrected CT secondary currents:

$$\begin{bmatrix} I_{A\_HV\_sec} \\ I_{B\_HV\_sec} \\ I_{C\_HV\_sec} \end{bmatrix} = \frac{1}{CTR_{HV}} \begin{bmatrix} I_{A\_HV} \\ I_{B\_HV} \\ 0 \end{bmatrix} \qquad \begin{bmatrix} I_{A\_LV\_sec} \\ I_{B\_LV\_sec} \\ I_{C\_LV\_sec} \end{bmatrix} = \frac{1}{CTR_{LV}} \begin{bmatrix} I_{A\_LV} \\ I_{B\_LV} \\ I_{C\_LV} \end{bmatrix}$$

The corrected CT secondary currents:

$$\begin{bmatrix} I_{A\_HV\_cor} \\ I_{B\_HV\_cor} \\ I_{C\_HV\_cor} \end{bmatrix} = \frac{1}{TAP_{HV}} MAT_{11} \begin{bmatrix} I_{A\_HV\_sec} \\ I_{B\_HV\_sec} \\ I_{C\_HV\_sec} \end{bmatrix} \qquad \begin{bmatrix} I_{A\_LV\_cor} \\ I_{B\_LV\_cor} \\ I_{C\_LV\_cor} \end{bmatrix} = \frac{1}{TAP_{LV}} MAT_{0} \begin{bmatrix} I_{A\_LV\_sec} \\ I_{B\_LV\_sec} \\ I_{C\_LV\_sec} \end{bmatrix}$$

The operating and restraint currents with a shorted CT:

Phase A:

$$I_{OPA} = \left| I_{A\_HV\_cor} + I_{A\_LV\_cor} \right| = 0.422$$
$$I_{RTA} = \left| I_{A\_HV\_cor} \right| + \left| I_{A\_LV\_cor} \right| = 1.154$$

Phase B:

$$I_{OPB} = \left| I_{B\_HV\_cor} + I_{B\_LV\_cor} \right| = 2.252 \times 10^{-5}$$
$$I_{RTB} = \left| I_{B\_HV\_cor} \right| + \left| I_{B\_LV\_cor} \right| = 1.464$$

Phase C:

$$I_{OPC} = \left| I_{C\_HV\_cor} + I_{C\_LV\_cor} \right| = 0.422$$
$$I_{RTC} = \left| I_{C\_HV\_cor} \right| + \left| I_{C\_LV\_cor} \right| = 1.154$$

For phases A and C,

$$\frac{I_{OPA}}{I_{RTA}} = \frac{I_{OPC}}{I_{RTC}} = 36.6\%$$

If the slope is set larger than 36.6%, no transformer differential element will operate for this condition. Otherwise, the phases A and C differential elements will operate.

9.4 LINE CURRENT DIFFERENTIAL PROTECTION

Similar to a bus or a transformer, a transmission line can also be protected by a differential protection scheme. The relay number for line current differential

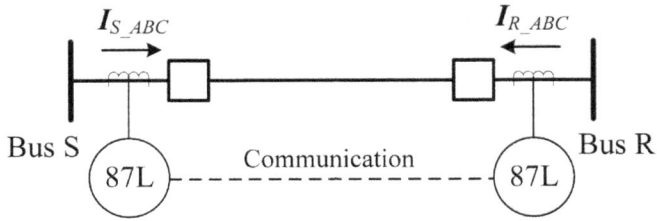

FIGURE 9.9 Line current differential protection diagram.

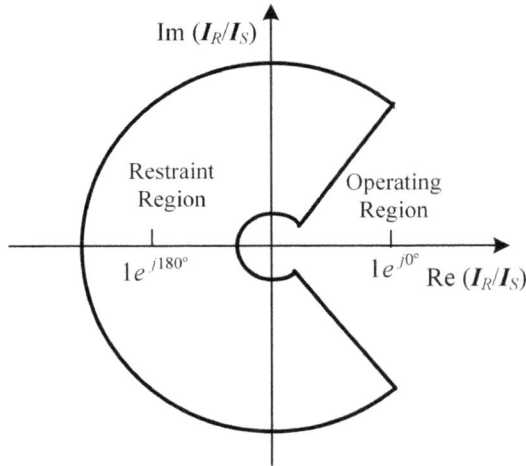

FIGURE 9.10 General principle of the alpha-plane-based method.

protection is 87L, as shown in Figure 9.9. When calculating the operating current I_{OP}, the transmission line charging current needs to be considered [9], as shown in Equation (9.6). The value of charging current can be estimated based on transmission line modeling and simulation.

$$I_{OP} = \left| I_S + I_R - I_{charging} \right| \tag{9.6}$$

9.4.1 Alpha-Plane-Based Method

Besides the traditional method, which calculates an operating current I_{OP} and a restraint current I_{RT}, an alternative method for the line current differential scheme based on the alpha plane can be used.

The general principle of the alpha-plane-based method can be illustrated using Figure 9.10. The ratio I_R / I_S for each phase (A, B, or C) is calculated as a complex number point, and it is mapped to a complex plane. If the point falls inside the restraint region, it indicates a load condition or external fault involving this phase. If the point falls outside the restraint region, it indicates an internal fault involving this phase. More detailed illustrations can be found in pertinent publications [10–12].

9.5 MATLAB IMPLEMENTATION

In this section, we will illustrate the implementation of line current differential (87L) protection functions using MATLAB. The functions are developed for line current differential relays located at the two ends of a transmission line. The voltage level of the transmission line is 230 kV line to line. The total length of the transmission line is 80 km. We simulate a phase-A-to-ground fault. The location of the fault is approximately 30 km from one end of the line. Implementation of the percentage differential method (low-impedance differential method) is presented first. Implementation of the alpha-plane-based method is presented later.

9.5.1 PERCENTAGE DIFFERENTIAL METHOD

The *Relay_main.m* is the master file. Running this file will run the entire line differential protection program. The MATLAB code of the *Relay_main.m* file is as follows:

```
%Relay_main.m, the master file of the program
%Begin
clear all;
close all;
clc;
%Read Bus S and Bus R current data
Relay_readdata;
%Configure the Relay settings
Relay_setting;
%Process the input currents through a filter
Relay_filter;
%Create phasors
Relay_phasor;
%The calculation and trip logics for line current
differential protection
Relay_87L;
%Plot the results
Relay_plot;
%End
```

The *Relay_readdata.m* reads the input currents from both ends of the transmission line. In this example, the input data were saved in a MATLAB file named "data_87L.mat". The data were obtained from a simulation case. The MATLAB code of the *Relay_readdata.m* file is shown below. The data are also resampled to 16 samples per cycle.

```
%Relay_readdata.m, reading the input currents
%Begin
CTR = 1.0;
PTR = 1.0;
VIread = load('data_87L.mat');
Isa = VIread.iBus02aBus04a;%Bus S phase A current
Isb = VIread.iBus02bBus04b;%Bus S phase B current
Isc = VIread.iBus02cBus04c;%Bus S phase C current
Ira = VIread.iBus06aBus6sa;%Bus R phase A current
Irb = VIread.iBus06bBus6sb;%Bus R phase B current
Irc = VIread.iBus06cBus6sc;%Bus R phase C current
time_before = VIread.t;
dt_before = time_before(2)-time_before(1);
RS_before = round( (1*3.0)/(60*dt_before) );
RS_after = 16*3;
%Resample the data to 16 samples per cycle
ISA = resample(Isa, RS_after, RS_before);
ISB = resample(Isb, RS_after, RS_before);
ISC = resample(Isc, RS_after, RS_before);
IRA = resample(Ira, RS_after, RS_before);
IRB = resample(Irb, RS_after, RS_before);
IRC = resample(Irc, RS_after, RS_before);
len = length(ISA);
%End
```

The *Relay_setting.m* configures the relay settings. In this example, we set the percentages of CT saturation as 0% for all the CTs. The values of line charging current and minimum current threshold are valid for this case only. Readers should use other values for other case studies. The MATLAB code of the *Relay_setting.m* file is shown as follows:

```
%Relay_setting.m, relay settings
%Begin
betas = 0.25;%The current differential tripping threshold
for BUS S
betar = 0.25;%The current differential tripping threshold
for BUS R
%CT saturation percentages for each phase
CTsAsat = 0;
CTsBsat = 0;
CTsCsat = 0;
CTrAsat = 0;
CTrBsat = 0;
CTrCsat = 0;
Icharge = 40;%The approximate line charging current
Imin = 10;%Minimum current threshold
%End
```

The *Relay_filter.m* performs a filter algorithm on the input currents. The MATLAB code of the *Relay_filter.m* file is shown as follows.

```
%Relay_filter.m
%Begin
RS = 16;
filter = 1;
%Filter, if (filter==0), Filter will be skipped
%{
ISA, ISB, ISC, IRA, IRB, IRC are the Bus S and Bus R
currents after being processed by the filter
%}
if (filter==1)
    Isa = zeros(len,1);
    Isb = zeros(len,1);
    Isc = zeros(len,1);
    Ira = zeros(len,1);
    Irb = zeros(len,1);
    Irc = zeros(len,1);
    for i = RS:len
        for k = 1:RS
            Isa(i) = Isa(i)+cos(2*pi*(k-1)/RS)*ISA(i-RS+k);
    end
        Isa(i) = Isa(i)*2/RS;

        for k = 1:RS
            Isb(i) = Isb(i)+cos(2*pi*(k-1)/RS)*ISB(i-RS+k);
        end
        Isb(i) = Isb(i)*2/RS;

        for k = 1:RS
        Isc(i) = Isc(i)+cos(2*pi*(k-1)/RS)*ISC(i-RS+k);
        end
        Isc(i) = Isc(i)*2/RS;

        for k = 1:RS
            Ira(i) = Ira(i)+cos(2*pi*(k-1)/RS)*IRA(i-RS+k);
        end
        Ira(i) = Ira(i)*2/RS;

        for k = 1:RS
            Irb(i) = Irb(i)+cos(2*pi*(k-1)/RS)*IRB(i-RS+k);
        end
        Irb(i) = Irb(i)*2/RS;

        for k = 1:RS
            Irc(i) = Irc(i)+cos(2*pi*(k-1)/RS)*IRC(i-RS+k);
        end
            Irc(i) = Irc(i)*2/RS;
    end
    ISA = Isa;
    ISB = Isb;
    ISC = Isc;
    IRA = Ira;
```

```
    IRB = Irb;
    IRC = Irc;
end
%End
```

The *Relay_phasor.m* calculates current phasors. The MATLAB code of the *Relay_phasor.m* file is as follows:

```
%Relay_phasor.m, phasor creation
%Begin
ISAcpx = zeros(1,len);
ISBcpx = zeros(1,len);
ISCcpx = zeros(1,len);
IRAcpx = zeros(1,len);
IRBcpx = zeros(1,len);
IRCcpx = zeros(1,len);
ISAsat = zeros(1,len);
ISBsat = zeros(1,len);
ISCsat = zeros(1,len);
IRAsat = zeros(1,len);
IRBsat = zeros(1,len);
IRCsat = zeros(1,len);
for v = (RS/4+1):len
    ISAcpx(v) = (ISA(v)+ 1i*ISA(v-RS/4))/sqrt(2);
    ISBcpx(v) = (ISB(v)+ 1i*ISB(v-RS/4))/sqrt(2);
    ISCcpx(v) = (ISC(v)+ 1i*ISC(v-RS/4))/sqrt(2);
    IRAcpx(v) = (IRA(v)+ 1i*IRA(v-RS/4))/sqrt(2);
    IRBcpx(v) = (IRB(v)+ 1i*IRB(v-RS/4))/sqrt(2);
    IRCcpx(v) = (IRC(v)+ 1i*IRC(v-RS/4))/sqrt(2);
    ISAsat(v) = (1-CTsAsat)*ISAcpx(v);
    ISBsat(v) = (1-CTsBsat)*ISBcpx(v);
    ISCsat(v) = (1-CTsCsat)*ISCcpx(v);
    IRAsat(v) = (1-CTrAsat)*IRAcpx(v);
    IRBsat(v) = (1-CTrBsat)*IRBcpx(v);
    IRCsat(v) = (1-CTrCsat)*IRCcpx(v);
end
%End
```

The *Relay_87L.m* is the main function for the line current differential protection. It performs calculation and checks trip logics. The MATLAB code of the *Relay_87L.m* file is shown as follows.

```
%Relay_87L.m
%The calculation and trip logics for line current
differential protection
%Begin
IS_OPA = zeros(1,len);
IS_OPB = zeros(1,len);
IS_OPC = zeros(1,len);
```

```
IR_OPA = zeros(1,len);
IR_OPB = zeros(1,len);
IR_OPC = zeros(1,len);
IS_RTA = zeros(1,len);
IS_RTB = zeros(1,len);
IS_RTC = zeros(1,len);
IR_RTA = zeros(1,len);
IR_RTB = zeros(1,len);
IR_RTC = zeros(1,len);
%Calculating the operating and restraint currents
for v = (RS/4+1):len
    IS_OPA(v) = abs(ISAsat(v)+IRAsat(v)-Icharge);
    IS_OPB(v) = abs(ISBsat(v)+IRBsat(v)-Icharge);
    IS_OPC(v) = abs(ISCsat(v)+IRCsat(v)-Icharge);
    IR_OPA(v) = abs(IRAsat(v)+ISAsat(v)-Icharge);
    IR_OPB(v) = abs(IRBsat(v)+ISBsat(v)-Icharge);
    IR_OPC(v) = abs(IRCsat(v)+ISCsat(v)-Icharge);

    IS_RTA(v) = abs(ISAsat(v))+abs(IRAsat(v));
    IS_RTB(v) = abs(ISBsat(v))+abs(IRBsat(v));
    IS_RTC(v) = abs(ISCsat(v))+abs(IRCsat(v));
    IR_RTA(v) = abs(IRAsat(v))+abs(ISAsat(v));
    IR_RTB(v) = abs(IRBsat(v))+abs(ISBsat(v));
    IR_RTC(v) = abs(IRCsat(v))+abs(ISCsat(v));
end
%The pick-up and trip logics
SA_pu = zeros(1,len); %Bus S Phase A pick-up logic;
SB_pu = zeros(1,len); %Bus S Phase B pick-up logic;
SC_pu = zeros(1,len); %Bus S Phase C pick-up logic;
S87L_trip = zeros(1,len); %Bus S overall trip logic
RA_pu = zeros(1,len); %Bus R Phase A pick-up logic;
RB_pu = zeros(1,len); %Bus R Phase B pick-up logic;
RC_pu = zeros(1,len); %Bus R Phase C pick-up logic;
R87L_trip = zeros(1,len); %Bus R overall trip logic
for i = 1:len
    if (IS_OPA(i)>betas*IS_RTA(i))&&(abs(ISAsat(i))>Imin)
        SA_pu(i) = 1;
    end
    if (IS_OPB(i)>betas*IS_RTB(i))&&(abs(ISBsat(i))>Imin)
        SB_pu(i) = 1;
    end
    if (IS_OPC(i)>betas*IS_RTC(i))&&(abs(ISCsat(i))>Imin)
        SC_pu(i) = 1;
    end
    S87L_trip(i) = SA_pu(i)||SB_pu(i)||SC_pu(i);

    if (IR_OPA(i)>betar*IR_RTA(i))&&(abs(IRAsat(i))>Imin)
        RA_pu(i) = 1;
    end
```

```
      if (IR_OPB(i)>betar*IR_RTB(i))&&(abs(IRBsat(i))>Imin)
          RB_pu(i) = 1;
      end
      if (IR_OPC(i)>betar*IR_RTC(i))&&(abs(IRCsat(i))>Imin)
          RC_pu(i) = 1;
      end
      R87L_trip(i) = RA_pu(i)||RB_pu(i)||RC_pu(i);
end
%End
```

The *Relay_plot.m* plots the results. The MATLAB code of the *Relay_plot.m* file is shown as follows.

```
%Relay_plot.m, plotting the results
%Begin
t = zeros(len,1);
freq = 60;
for i = 1:len
    t(i) = (i-1)*1000.0/(freq*RS);
end

figure;
plot(t,IS_OPA,'--b',t,IS_RTA,'r');
xlabel('Time (ms)');
ylabel('Current (A)');
legend('ISOPA','ISRTA');

figure;
plot(t,IR_OPA,'--b',t,IR_RTA,'r');
xlabel('Time (ms)');
ylabel('Current (A)');
legend('IROPA','IRRTA');

figure;
plot(t,S87L_trip,'k');
xlabel('Time (ms)');
ylabel('Logic Value');
ylim([-0.1 1.1]);

figure;
plot(t,R87L_trip,'k');
xlabel('Time (ms)');
ylabel('Logic Value');
ylim([-0.1 1.1]);
%End
```

The phase A operating and restraint currents (IOP and IRT) calculated by the relays at Bus S and Bus R are shown in Figure 9.11.

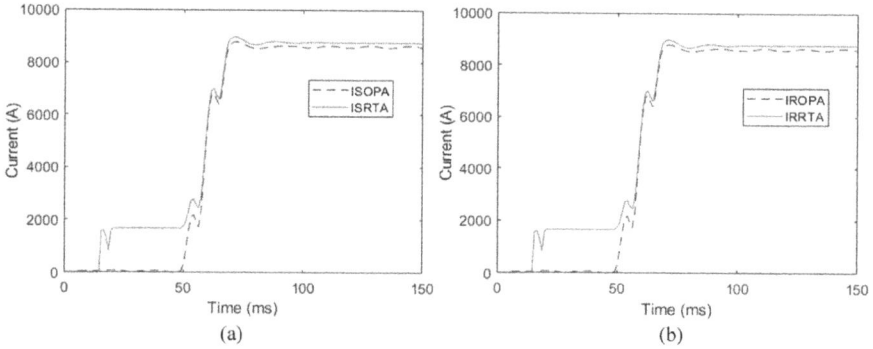

FIGURE 9.11   Phase A operating and restraint currents at (a) Bus S and (b) Bus R.

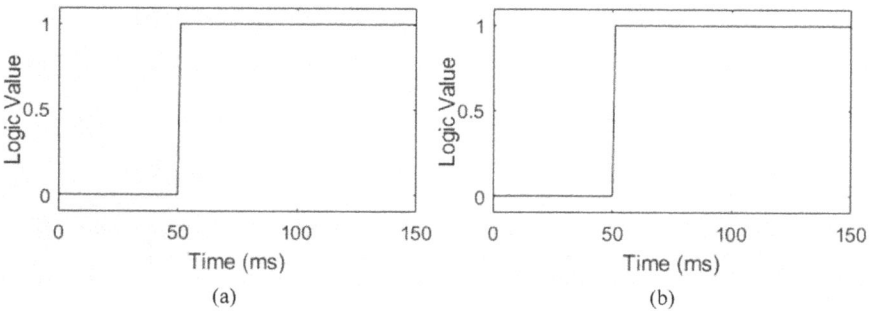

FIGURE 9.12   Relay responses at (a) Bus S and (b) Bus R.

The relay responses (trip logic value) at Bus S and Bus R are shown in Figure 9.12.

9.5.2 ALPHA-PLANE-BASED METHOD

The *Relay_main.m* is the master file. Running this file will run the entire line differential protection program. The MATLAB code of the *Relay_main.m* file is shown as follows:

```
%Relay_main.m, the master file of the program
%Begin
clear all;
close all;
clc;
%Read Bus S and Bus R current data
Relay_readdata;
%Configure the Relay settings
```

```
Relay_setting;
%Process the input currents through a filter
Relay_filter;
%Create phasors
Relay_phasor;
%Calculation for line current differential protection using
Alpha plane
Relay_87L_Alpha;
%Plot the results
Relay_plot;
%End
```

The *Relay_readdata.m* reads the input currents from both ends of the transmission line. In this example, the input data were saved in a MATLAB file named "data_87L.mat". The data were obtained from a simulation case. The MATLAB code of the *Relay_readdata.m* file is shown below. The data are also resampled to 16 samples per cycle.

```
%Relay_readdata.m, reading the input currents
%Begin
CTR = 1.0;
PTR = 1.0;
VIread = load('data_87L.mat');
Isa = VIread.iBus02aBus04a;%Bus S phase A current
Isb = VIread.iBus02bBus04b;%Bus S phase B current
Isc = VIread.iBus02cBus04c;%Bus S phase C current
Ira = VIread.iBus06aBus6sa;%Bus R phase A current
Irb = VIread.iBus06bBus6sb;%Bus R phase B current
Irc = VIread.iBus06cBus6sc;%Bus R phase C current
time_before = VIread.t;
dt_before = time_before(2)-time_before(1);
RS_before = round( (1*3.0)/(60*dt_before) );
RS_after = 16*3;
%Resample the data to 16 samples per cycle
ISA = resample(Isa, RS_after, RS_before);
ISB = resample(Isb, RS_after, RS_before);
ISC = resample(Isc, RS_after, RS_before);
IRA = resample(Ira, RS_after, RS_before);
IRB = resample(Irb, RS_after, RS_before);
IRC = resample(Irc, RS_after, RS_before);
len = length(ISA);
%End
```

The *Relay_setting.m* configures the relay settings. In this example, we set the percentages of CT saturation as 0% for all the CTs. We also set the two angles of the restraint region to be 85 degrees and 275 degrees, respectively. The values of line charging current and minimum current threshold are valid for this case only. Readers should use other values for other case studies. The MATLAB code of the *Relay_setting.m* file is as follows:

```
%Relay_setting.m, relay settings
%Begin
%CT saturation percentages for each phase
CTsAsat = 0;
CTsBsat = 0;
CTsCsat = 0;
CTrAsat = 0;
CTrBsat = 0;
CTrCsat = 0;
Icharge = 40;%The approximate line charging current
Imin = 10;%Minimum current threshold
Outer_R = 3;%Outer radius of the circle;
Inner_R = 1.0/Outer_R;%Inner radius of the circle;
Angle1 = 85*pi/180;
Angle2 = 275*pi/180;
%End
```

The *Relay_filter.m* performs a filter algorithm on the input currents. The MATLAB code of the *Relay_filter.m* file is shown as follows:

```
%Relay_filter.m
%Begin
RS = 16;
filter = 1;
%Filter, if (filter==0), Filter will be skipped
%{
ISA, ISB, ISC, IRA, IRB, IRC are the Bus S and Bus R
currents after being processed by the filter
%}
if (filter==1)
    Isa = zeros(len,1);
    Isb = zeros(len,1);
    Isc = zeros(len,1);
    Ira = zeros(len,1);
    Irb = zeros(len,1);
    Irc = zeros(len,1);
    for i = RS:len
        for k = 1:RS
            Isa(i) = Isa(i)+cos(2*pi*(k-1)/RS)*ISA(i-RS+k);
        end
        Isa(i) = Isa(i)*2/RS;

        for k = 1:RS
            Isb(i) = Isb(i)+cos(2*pi*(k-1)/RS)*ISB(i-RS+k);
        end
        Isb(i) = Isb(i)*2/RS;

        for k = 1:RS
            Isc(i) = Isc(i)+cos(2*pi*(k-1)/RS)*ISC(i-RS+k);
        end
```

```
        Isc(i) = Isc(i)*2/RS;

        for k = 1:RS
            Ira(i) = Ira(i)+cos(2*pi*(k-1)/RS)*IRA(i-RS+k);
    end
        Ira(i) = Ira(i)*2/RS;

        for k = 1:RS
            Irb(i) = Irb(i)+cos(2*pi*(k-1)/RS)*IRB(i-RS+k);
        end
        Irb(i) = Irb(i)*2/RS;


        for k = 1:RS
            Irc(i) = Irc(i)+cos(2*pi*(k-1)/RS)*IRC(i-RS+k);
        end
        Irc(i) = Irc(i)*2/RS;
    end
    ISA = Isa;
    ISB = Isb;
    ISC = Isc;
    IRA = Ira;
    IRB = Irb;
    IRC = Irc;
end
%End
```

The *Relay_phasor.m* calculates current phasors. The MATLAB code of the *Relay_phasor.m* file is as follows:

```
%Relay_phasor.m, phasor creation
%Begin
ISAcpx = zeros(1,len);
ISBcpx = zeros(1,len);
ISCcpx = zeros(1,len);
IRAcpx = zeros(1,len);
IRBcpx = zeros(1,len);
IRCcpx = zeros(1,len);
ISAsat = zeros(1,len);
ISBsat = zeros(1,len);
ISCsat = zeros(1,len);
IRAsat = zeros(1,len);
IRBsat = zeros(1,len);
IRCsat = zeros(1,len);
for v = (RS/4+1):len
    ISAcpx(v) = (ISA(v)+ 1i*ISA(v-RS/4))/sqrt(2);
    ISBcpx(v) = (ISB(v)+ 1i*ISB(v-RS/4))/sqrt(2);
    ISCcpx(v) = (ISC(v)+ 1i*ISC(v-RS/4))/sqrt(2);
    IRAcpx(v) = (IRA(v)+ 1i*IRA(v-RS/4))/sqrt(2);
```

```
    IRBcpx(v)  =  (IRB(v)+ 1i*IRB(v-RS/4))/sqrt(2);
    IRCcpx(v)  =  (IRC(v)+ 1i*IRC(v-RS/4))/sqrt(2);
    ISAsat(v)  =  (1-CTsAsat)*ISAcpx(v);
    ISBsat(v)  =  (1-CTsBsat)*ISBcpx(v);
    ISCsat(v)  =  (1-CTsCsat)*ISCcpx(v);
    IRAsat(v)  =  (1-CTrAsat)*IRAcpx(v);
    IRBsat(v)  =  (1-CTrBsat)*IRBcpx(v);
    IRCsat(v)  =  (1-CTrCsat)*IRCcpx(v);
end
%End
```

The *Relay_87L_Alpha.m* is the main function for the line current differential protection. It performs alpha-plane-based calculation. The trip logics are not included in this example. The MATLAB code of the *Relay_87L_Alpha.m* file is as follows:

```
%Relay_87L_Alpha.m
%The calculation for line current differential protection
%Begin
k_A = zeros(1,len);
k_B = zeros(1,len);
k_C = zeros(1,len);

%Calculating the operating and restraint currents
for v = 1:RS
    k_A(v)  = -1;
    k_B(v)  = -1;
    k_C(v)  = -1;
end
for v = (RS+1):len
    k_A(v)  = IRAsat(v)/ISAsat(v);
    k_B(v)  = IRBsat(v)/ISBsat(v);
    k_C(v)  = IRCsat(v)/ISCsat(v);
end
%End
```

The *Relay_plot.m* plots the results. The MATLAB code of the *Relay_plot.m* file is as follows:

```
%Relay_plot.m, plotting the results
%Begin
figure;
th = Angle1:pi/200:Angle2;
xunit = Outer_R*cos(th);
yunit = Outer_R*sin(th);
plot(xunit, yunit,'m','LineWidth',1.2);%Plot the outer
circle
hold on;
xunit = Inner_R*cos(th);
```

```
yunit = Inner_R*sin(th);
plot(xunit, yunit,'m','LineWidth',1.2);%Plot the inner
circle
x_axis(1) = Inner_R*cos(Angle1);
y_axis(1) = Inner_R*sin(Angle1);
x_axis(2) = Outer_R*cos(Angle1);
y_axis(2) = Outer_R*sin(Angle1);
plot(x_axis,y_axis,'m','LineWidth',1.2); %Plot the upper
boundary
x_axis(1) = Inner_R*cos(Angle2);
y_axis(1) = Inner_R*sin(Angle2);
x_axis(2) = Outer_R*cos(Angle2);
y_axis(2) = Outer_R*sin(Angle2);
plot(x_axis,y_axis,'m','LineWidth',1.2); %Plot the lower
boundary
x_axis = zeros(1,2);
y_axis = zeros(1,2);
x_axis(1) = -3.5;
x_axis(2) = 3.5;
plot(x_axis,y_axis,'k','LineWidth',0.5); %Plot the
horizontal axis
x_axis = zeros(1,2);
y_axis = zeros(1,2);
y_axis(1) = -3.5;
y_axis(2) = 3.5;
plot(x_axis,y_axis,'k','LineWidth',0.5); %Plot the verticle
axis

hpA = plot(real(k_A),imag(k_A),'bx');
hpB = plot(real(k_B),imag(k_B),'r*');
hpC = plot(real(k_C),imag(k_C),'ko');
legend([hpA hpB hpC],'Phase A','Phase B','Phase C');
xlabel('Re (IR/IS)');
ylabel('Imag (IR/IS)');
xlim([-3.5 3.5]);
ylim([-3.5 3.5]);
%End
```

The results are shown in Figure 9.13. Because this is a phase-A-to-ground fault, only the phase A element response enters the operating region.

9.6   SUMMARY

In this chapter, we have mainly illustrated the principles of bus, transformer, and transmission line differential protection schemes. The MATLAB programming details of transmission line differential protection have been illustrated. Readers are encouraged to complete the trip logics for the alpha-plane-based method and implement the algorithms on their own computers to have a better understanding of relay functions and logics.

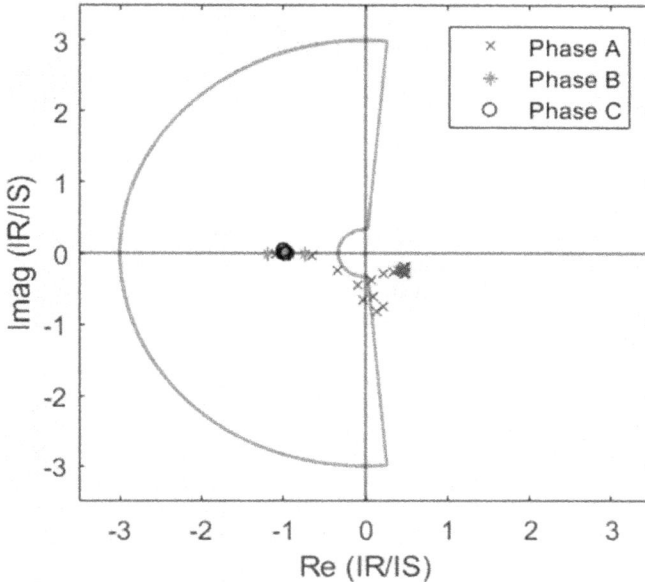

FIGURE 9.13 Results of the alpha-plane-based method.

9.7 PROBLEMS

Problem 9.1

Complete the MATLAB code of the trip logics for the alpha-plane-based method that has been illustrated in Section 9.5.

Problem 9.2

Compare line current differential protection with distance protection for transmission lines, briefly summarize their advantages and limitations.

BIBLIOGRAPHY

[1] J. L. Blackburn and T. J. Domin, *Protective Relaying: Principles and Applications*, 3rd Ed. CRC Press, 2006.

[2] P. M. Anderson, C. Henville, R. Rifaat, B. Johnson, and S. Meliopoulos, *Power System Protection*, 2nd Ed. Wiley, 2022.

[3] H. Lei and C. Singh, "Power system reliability evaluation considering cyber-malfunctions in substations," *Electric Power Systems Research*, vol. 129, pp. 160–169, December 2015.

[4] H. Lei and C. Singh, "Incorporating protection systems into composite power system reliability assessment," in IEEE Power and Energy Society General Meeting, July 2015, pp. 1–5, DOI: 10.1109/PESGM.2015.7285636.

[5] J. G. Andrichak and J. Cardenas, "Bus differential protection," in 22nd Western Protective Relay Conference, pp. 1–11. 1995.

[6] S. Kucuksari and G. G. Karady, "Experimental comparison of conventional and optical current transformers," *IEEE Transactions on Power Delivery*, vol. 25, no. 4, pp. 2455–2463, 2010.

[7] K. Behrendt, D. Costello, and S. E. Zocholl, "Considerations for using high-impedance or low-impedance relays for bus differential protection," in 2010 63rd Annual Conference for Protective Relay Engineers, pp. 1–15, 2010.

[8] C37.91-2021, IEEE Guide for Protecting Power Transformers, DOI: 10.1109/IEEE STD.2021.9471045.

[9] J. Bell, A. Hargrave, G. Smelich, and B. Smyth, "Considerations when using charging current compensation in line current differential applications," in Proceedings of the 72nd Annual Conference for Protective Relay Engineers, pp. 1–11, 2019.

[10] G. Benmouyal, "The trajectories of line current differential faults in the alpha plane," in Proceedings the 32nd Annual Western Protective Relay Conference, Spokane, WA, 2005.

[11] D. A. Tziouvaras, H. Altuve, G. Benmouyal, and J. Roberts, "Line differential protection with an enhanced characteristic," in 3rd Mediterranean Conference on Power Generation, Transmission, Distribution and Energy Conversion, 2002.

[12] H. Miller, J. Burger, N. Fischer, and B. Kasztenny, "Modern line current differential protection solutions," in 2010 63rd Annual Conference for Protective Relay Engineers, 2010, DOI: 10.1109/CPRE.2010.5469504.

10 Time-Domain Protection

The overcurrent, distance, and differential protection schemes illustrated in previous chapters are based on phasors. The phasors of currents and/or voltages are calculated before performing protection algorithms. Unlike phasor-based schemes, some protection schemes directly use the instantaneous values of currents and/or voltages for protection algorithms. The instantaneous values are also called time-domain quantities. Utilizing time-domain quantities instead of phasors could reduce processing time and achieve fast tripping. In this chapter, we illustrate the principles of two types of time-domain protection schemes: traveling wave-based protection and superimposed quantity-based protection. By focusing on the transient fault signatures, time-domain protection schemes offer improved selectivity, sensitivity, and security in fault detection.

10.1 TRAVELING WAVE-BASED PROTECTION INTRODUCTION

Traveling wave-based protection leverages the high-frequency electromagnetic transients that propagate through the transmission line at nearly the speed of light, enabling very fast fault detection and clearing [1, 2]. The general principle of traveling wave-based protection is shown in Figure 10.1. It is based on the fact that transient voltage and current traveling waves will be launched when a fault occurs on a transmission line [3]. The transient voltage and current traveling waves will propagate along the transmission line at approximately the speed of light. When the induced traveling waves arrive at the ends of the transmission line and are measured by the protective relays, the fault can be identified and cleared accordingly [4, 5].

The magnitude of the initial voltage traveling wave is determined by the abrupt voltage change at the fault occurrence instant, which depends on the fault inception angle (i.e., point on the wave of fault occurrence). The magnitude of the initial current traveling wave can be calculated using Equation (10.1), where v_F is the magnitude of the initial voltage traveling wave, i_F is the magnitude of the initial current traveling wave, and Z_C is the transmission line surge impedance.

$$i_F = \frac{v_F}{Z_c} \tag{10.1}$$

After traveling waves are launched by a fault on a transmission line, they will largely maintain their shapes and amplitudes when propagating along the transmission line until hitting a discontinuity point. A discontinuity point is a point at

DOI: 10.1201/9781003629481-10 **129**

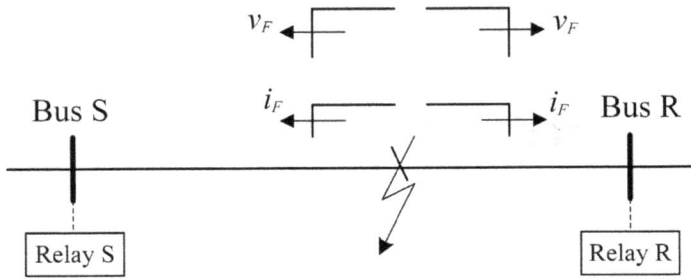

FIGURE 10.1   Voltage and current traveling waves launched by a fault.

which two or more branches (e.g., transmission lines and transformers) are connected. Typically, the terminals of a transmission line (such as Bus S and Bus R in Figure 10.1) are discontinuity points. Reflection occurs when a traveling wave hits a discontinuity point because of the mismatch between the terminating impedance and the characteristic impedance of the transmission line. The terminating impedance is the overall characteristic impedance of other branches connected to the discontinuity point. The measured traveling wave by the relay at a discontinuity point is the summation of the initial wave and the reflected wave because the relay cannot distinguish between the initial wave and the reflected wave.

The amplitude and polarity of the reflected current traveling wave can be calculated using Equation (10.2). The i_{RFT} in Equation (10.2) is the reflected current traveling wave, and the i_I is the incoming current traveling wave. Z_T is the terminating impedance, and Z_C is the transmission line characteristic impedance.

$$i_{RFT} = \frac{Z_C - Z_T}{Z_C + Z_T} i_I \qquad (10.2)$$

For example, if an ideal voltage source is connected to the left side of Bus S, the equivalent terminating impedance is zero. The reflected wave amplitude equals the incoming wave amplitude, with the same polarity. Thus, when an initial current traveling wave propagating on the transmission line hits Bus S, its amplitude will be doubled because the overall wave is the summation of the incoming wave and the reflected wave. If an open circuit is connected to the left side of Bus S, the equivalent terminating impedance is infinite. The reflected wave amplitude equals the incoming wave amplitude, with an opposite polarity. Thus, when an initial current traveling wave propagating on the transmission line hits Bus S, its amplitude will become zero.

Figure 10.2 shows the obtained current traveling waves at the fault point and the transmission line terminal from a simulation case with zero terminating impedance. The vertical axis is current amplitude, and the horizontal axis is time. We can observe that there is a time difference between the first steps (rising edges) of the two waveforms, which corresponds to the time it takes for the traveling wave propagating from the fault point to the line terminal. We can also observe that

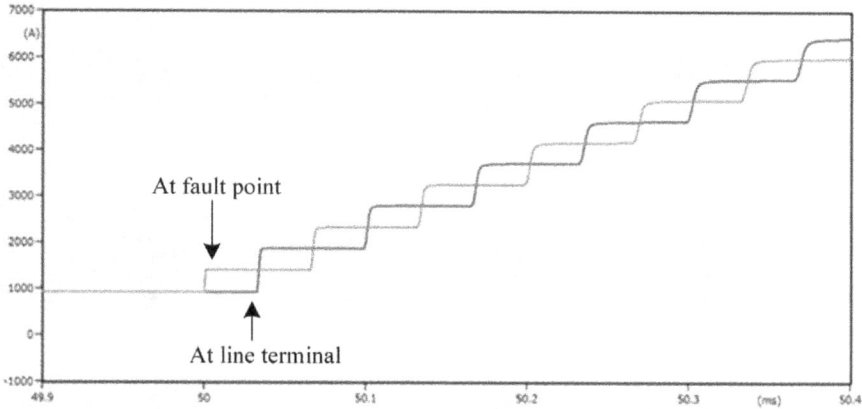

FIGURE 10.2   Current traveling wave reflection with zero terminating impedance.

the amplitude (height) of the first wave step at the line terminal is approximately twice that of the first wave step at the fault point. This is because the first wave step at the line terminal is the combination of the arriving wave and the reflected wave. From an instrumentation perspective, a relay is not able to differentiate the reflected wave from the arriving wave.

It should be noted that Figure 10.2 is an example from simulation. In actual power systems, we can only measure traveling waves at line terminals, and we are not able to directly measure traveling waves at a fault point. It should also be noted that in fault simulation, we will only be able to capture traveling wave features if a traveling wave model (such as Bergeron model or JMarti model) is used to represent the transmission line [6]. If the transmission line is represented by a lumped π or lumped RL model, the traveling wave features (the wave steps) exhibited in the transient response in actual transmission lines will be lost in the simulation.

10.2   LOCATING A FAULT BASED ON CURRENT TRAVELING WAVES

Figure 10.2 is an ideal case in which the terminating impedance equals zero, and thus the edges of the traveling wave steps are almost vertical. If other branches are connected, the terminating impedance will no longer be zero. For example, Figure 10.3 (a) shows a power system for simulation. The length of the transmission line connecting Bus S and Bus R is 82 km. There are two parallel short lines connecting Bus A to Bus S, both with a length of 18 km. A phase-A-to-ground fault occurs on the 82-km transmission line at approximately 0.033 s. The fault point F is approximately 12 km from Bus S. The phase A current traveling wave acquired at Bus S is shown in Figure 10.3 (b). The first wave step is the initial wave coming from point F. The second wave step includes the reflected wave

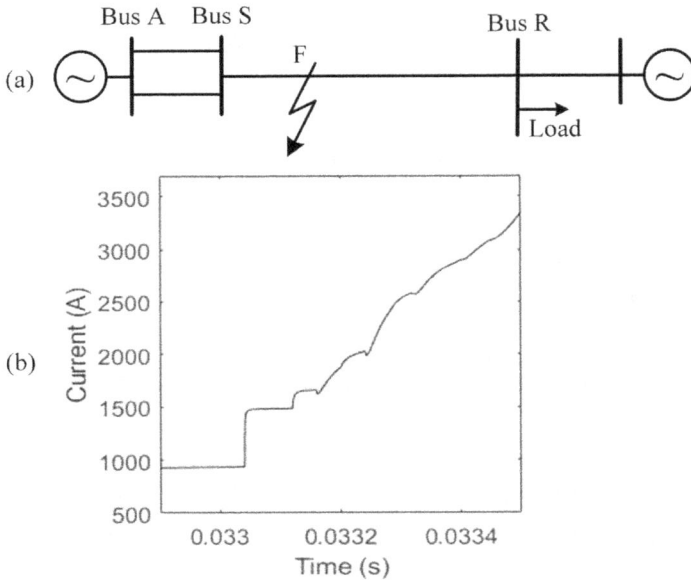

FIGURE 10.3 (a) A simulation system and (b) phase A current traveling wave acquired at Bus S.

from Bus S to point F and then returns to Bus S. The later wave steps involve the reflection at Bus A, which is associated with the equivalent source impedance. We can see that the shapes of the later wave steps are different from the first two wave steps.

The principle of traveling wave-based protection is to make protection decisions based on the successful recognition of the first 1–3 traveling wave steps. Current traveling waves are typically used in this process. Raw traveling waves are typically processed through anti-alias filters and then used for wave step recognition [7–9]. By identifying the time instants corresponding to the rising edges of current traveling wave steps, the fault occurrence time instants and location of the fault can be determined.

Fault currents consist of line-mode and ground-mode components. In implementation [5, 10], relays can perform a Clarke transformation to convert the acquired A, B, and C quantities (currents or voltages) to two line modes and a ground mode quantities. Then, use the converted quantities for rising-edge recognition. Equation (10.3) can be used to calculate the Clarke components with respect to phase A. The i_A, i_B, and i_C represent three-phase instantaneous currents. The two line modes are labeled as α and β, and the ground mode is labeled as 0. The equations for Clarke transformation with respect to phases B and C are provided in references [5, 10].

$$\begin{bmatrix} i_A^{\alpha} \\ i_A^{\beta} \\ i_A^{0} \end{bmatrix} = \frac{1}{3} \begin{bmatrix} 2 & -1 & -1 \\ 0 & \sqrt{3} & -\sqrt{3} \\ 1 & 1 & 1 \end{bmatrix} \begin{bmatrix} i_A \\ i_B \\ i_C \end{bmatrix} \tag{10.3}$$

Since the current traveling wave shown in Figure 10.3 (b) is obtained from a phase-A-to-ground fault, we can use Equation (10.3) to extract the line- and ground-mode components with respect to phase A. After performing the Clarke transformation, the α-line-mode component (i_A^{α}) will be used for rising-edge recognition.

To recognize the rising edges in the i_A^{α} waveform, high-frequency noises will first be filtered out by using a signal smoother. Then, the smoothened waveform will go through a differentiator to obtain the waveform of the derivative $d i_A^{\alpha} / dt$. A positive peak in the derivative waveform can be identified by comparing an instantaneous value with a threshold. If the instantaneous value is greater than the threshold, a positive peak in the $d i_A^{\alpha} / dt$ waveform is recognized, which means a rising edge in the i_A^{α} is recognized. The derivative waveform, $d i_A^{\alpha} / dt$, which is obtained based on the Figure 10.3 (b) waveform, is shown in Figure 10.4. It can be seen that the first two wave steps occur at approximately 33.04 ms and 33.12 ms, respectively.

If Bus S is the only terminal of the transmission line at which traveling wave measurements are available, then the arrival time instants of at least two traveling wave steps are needed to locate the fault. If traveling wave measurements from both Bus S and Bus R are available, then using the arrival time instants of the first traveling wave steps at both terminals is able to locate the fault.

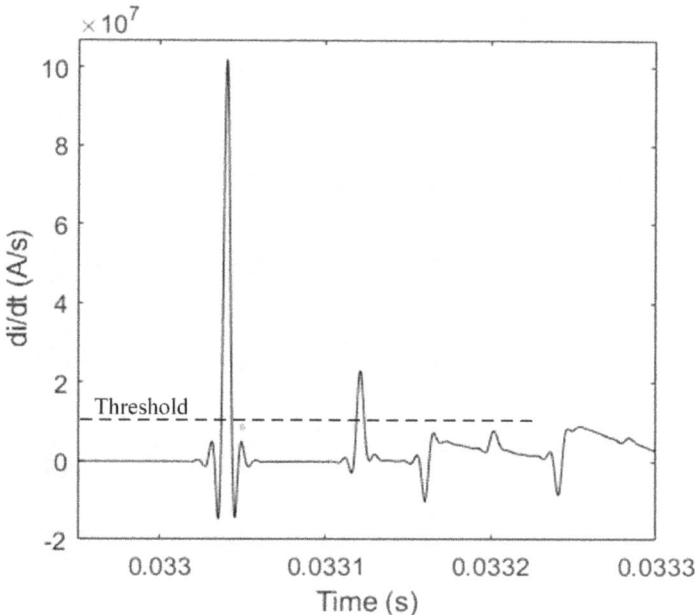

FIGURE 10.4 Derivative of the α-line-mode component of phase A current.

10.3 FAULT DIRECTION DETERMINATION

A torque-based method can be used to determine the direction of a fault (forward or reverse) [4]. The torque-based method computes the product of current and voltage traveling waves and determines the fault direction based on the polarity. The principle is similar to the torque-based method illustrated in Chapter 5.

The flow chart for implementing the torque-based method is shown in Figure 10.5. Phase A current and voltage are used in this flow chart because it uses a phase-A-to-ground fault as an example. Detailed results from an implementation are provided in reference [5].

If traveling wave measurements from both Bus S and Bus R are available, a traveling wave-based differential (TW87) protection scheme can be designed and implemented. The general principle is similar to the differential protection schemes illustrated in Chapter 9. Detailed examples can be found in references [11, 12].

10.4 SUPERIMPOSED QUANTITY-BASED PROTECTION

Superimposed quantities are also called incremental quantities or delta quantities [10]. The principle of superimposed quantities is based on comparing a pre-fault network with a faulted network [13–15]. For illustration purposes, a power system with double sources is shown in Figure 10.6. The double sources represent the equivalent voltages of the systems connected to the two ends of the line. In some radial distribution line cases, when one end of the line is connected, only one equivalent source is needed, and the problem can be even simplified. We use double sources for illustration because this situation has better generality.

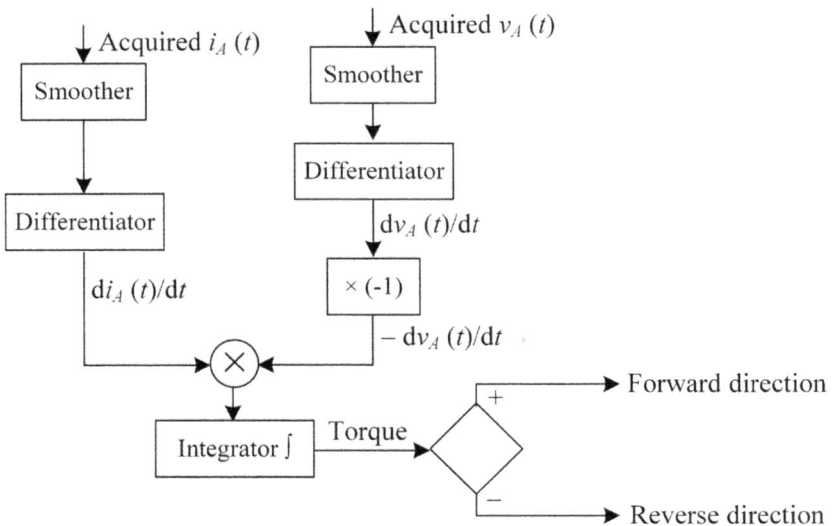

FIGURE 10.5 Procedures for determining fault direction.

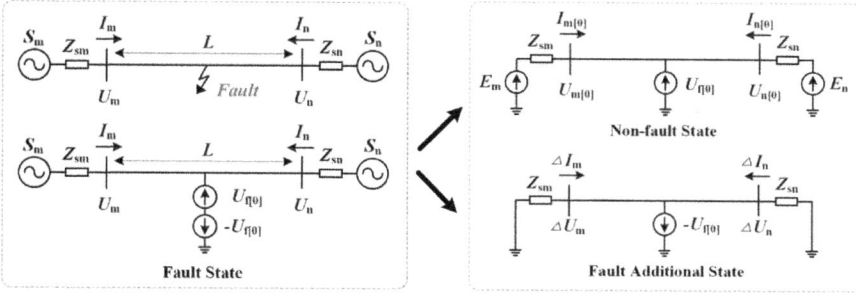

FIGURE 10.6 Decomposition of a power system fault state.

When a fault event occurs in the system, the overall fault state can be decoupled into a non-fault state and a fault additional state according to the superposition principle, as shown in Figure 10.6. The fault additional voltage $-U_{f[0]}$ can be represented by:

$$-U_{f[0]} = -\Delta I_m \left(Z_{sm} + Z_{location} \right) \tag{10.4}$$

The superimposed voltage ΔU_m measured by the device on the m side is as follows:

$$\Delta U_m = -\Delta I_m Z_{sm} \tag{10.5}$$

In Equations (10.4) and (10.5), Z_{sm} is the equivalent source impedance for S_m. ΔI_m is the superimposed current measured by the device at the m side. $Z_{location}$ is the impedance computed by the device at the m side based on measurements, and it is the indicator of fault location. By taking measurements also at the other side of the line, two more similar equations can be obtained and the effect of fault path resistance (i.e., the resistance between the fault point and the ground) can be addressed. Based on the superimposed voltages and currents, the fault location ($Z_{location}$) can be determined.

The superimposed quantities in the phasor domain are adequate for the cases in which the response time is not critical. To achieve fast tripping, time-domain-based measurements are needed to capture the dynamic and transient features. In the time domain, the superimposed voltage $\Delta u_k(t)$ and current $\Delta i_k(t)$ quantities for each phase are computed by subtracting the previously stored voltage $u_k(t-T)$ and current $i_k(t-T)$ values from the presently sampled voltage $u_k(t)$ and current $i_k(t)$ values, as shown in Equations (10.6) and (10.7). T is the length of a cycle ($T = 16.67$ ms). The subscript k represents the measurements taken at different ends of a line.

$$\Delta u_k(t) = u_k(t) - u_k(t-T) \tag{10.6}$$

$$\Delta i_k\left(t\right)=i_k\left(t\right)-i_k\left(t-T\right) \tag{10.7}$$

By using the superimposed currents and voltages, various protection schemes can be designed and implemented [10, 13–20]. For example, authors in [10, 13, 20] have implemented transmission line directional elements based on superimposed quantities in commercial relays to improve relay response speed. Some of these superimposed quantity-based schemes calculate phasors using either half-cycle or full-cycle digital filters. Some of these schemes make decisions based on instantaneous quantities, in which calculations are performed and updated over a moving window.

Some commercial relays have utilized superimposed quantities in directional elements to achieve ultra-high-speed response [13, 16]. The principle is similar to the torque-based elements illustrated in Chapter 5. The directional element uses both superimposed voltage and current quantities to calculate instantaneous superimposed torque quantities in the time domain. The direction determination of faults is based on the relative polarities of voltages and currents. Detailed implementation of superimposed quantity-based protection schemes in power systems is included in references [10, 13–20].

10.5 SUMMARY

In this chapter, protection schemes based on traveling waves and superimposed quantities have been introduced. Traveling wave-based protection detects faults by analyzing the traveling waves that propagate along transmission lines after fault occurrence. It can achieve ultra-high-speed fault detection but requires sophisticated high-frequency measurement and analysis techniques. Superimposed quantity-based protection analyzes incremental differences in voltage and current quantities to detect and identify faults. Although generally not as fast as traveling wave-based protection, it can still achieve relatively high-speed fault detection, depending on signal processing algorithms. Superimposed quantity-based protection schemes are generally more affordable than traveling wave-based schemes due to lower hardware requirements.

BIBLIOGRAPHY

[1] X. Dong, W. Kong, and T. Cui, "Fault classification and faulted-phase selection based on the initial current traveling wave," *IEEE Transactions on Power Delivery*, vol. 24, no. 2, pp. 552–559, 2009.

[2] X. Dong, Y. Ge, and J. He, "Surge impedance relay," *IEEE Transactions on Power Delivery*, vol. 20, no. 2, pp. 1247–1256, 2005.

[3] H. Lei, J. Geng, and B. K. Johnson, "Influence of superconducting fault current limiters on traveling wave based protection," *IEEE Transactions on Applied Superconductivity*, vol. 29, no. 5, August 2019, Article No. 5602305.

[4] E. O. Schweitzer, A. Guzman, M. V. Mynam, V. Skendzic, B. Kasztenny, and S. Marx, "Locating faults by the traveling waves they launch," in Proceeding of the 67th Annual Conference for Protective Relay Engineers, pp. 95–110, 2014.

[5] H. Lei, J. Gui, and B. K. Johnson, "Impact of saturated iron core superconducting fault current limiters on traveling-wave-based protection," *IEEE Transactions on Applied Superconductivity*, vol. 33, no. 8, November 2023, Article No. 5601308.

[6] J. R. Marti, "Accurate modelling of frequency-dependent transmission lines in electromagnetic transient simulations," *IEEE Transactions on Power Apparatus and Systems*, vol. PAS-101, no. 1, pp. 147–157, 1982.

[7] E. O. Schweitzer, A. Guzman, M. V. Mynam, V. Skendzic, B. Kasztenny, and S. Marx, "Protective relays with traveling wave technology revolutionize fault locating," *IEEE Power and Energy Magazine*, vol. 14, no. 2, pp. 114–120, 2016.

[8] A. Guzman, M. V. Mynam, V. Skendzic, J. L. Eternod, and R. M. Morales, "Traveling-wave and incremental quantity directional elements speed up directional comparison protection schemes," in Proceedings of the 14th International Conference on Developments in Power System Protection, pp. 1–6, 2018.

[9] S. Marx, B. K. Johnson, A. Guzman, V. Skendzic, and M. V. Mynam, "Traveling wave fault location in protective relays: design, testing, and results," in Proceedings of the 16th Annual Georgia Tech Fault and Disturbance Analysis Conference, Atlanta, GA, pp. 1–14, 2013.

[10] B. Kasztenny, A. Guzman, N. Fischer, M. V. Mynam, and D. Taylor, "Practical setting considerations for protective relays that use incremental quantities and traveling waves," in Proceedings of the 43rd Annual Western Protective Relay Conference, Spokane, WA, pp. 1–25, 2016.

[11] A. Lei, X. Dong, S. Shi, B. Wang, and V. Terzija, "Equivalent traveling waves based current differential protection of EHV/UHV transmission lines," *International Journal of Electrical Power & Energy Systems*, vol. 97, pp. 282–289, 2018.

[12] L. Tang, X. Dong, S. Luo, S. Shi, and B. Wang, "A new differential protection of transmission line based on equivalent travelling wave," *IEEE Transactions on Power Delivery*, vol. 32, no. 3, pp. 1359–1369, 2016.

[13] E. O. Schweitzer, III, B. Kasztenny, A. Guzmán, V. Skendzic, and M. V. Mynam, "Speed of line protection—can we break free of phasor limitations?" in Proceedings of the 41st Annual Western Protective Relay Conference, Spokane, WA, October 2014.

[14] H. Lei, H. Samkari, Y. Chakhchoukh, J. Geng, and B. K. Johnson, "Impact of resistive SFCLs on superimposed quantities in power system faults," *IEEE Transactions on Applied Superconductivity*, vol. 31, no. 7, pp. 1–8, 2021.

[15] A. Sirisha and S. R. Bhide, "Incremental quantities based relays," in 2014 IEEE International Conference on Power, Automation and Communication (INPAC), pp. 27–32, 2014.

[16] G. Benmouyal and J. Roberts, "Superimposed quantities: their true nature and application in relays," in Proceedings of the 26th Annual Western Protective Relay Conference, Spokane, WA, 1999.

[17] H. S. Samkari and B. K. Johnson, "Impact of distributed inverter-based resources on incremental quantities-based protection," in 2021 IEEE Power & Energy Society General Meeting (PESGM), pp. 1–5, 2021.

[18] J. Gui, H. Lei, and B. K. Johnson, "A superimposed quantity-based protection method for power systems with inverter-based resources," *Electric Power Systems Research*, vol. 232, 2024, Article No. 110390.

[19] A. Guzman, C. Labuschagne, and B. L. Qin, "Reliable busbar and breaker failure protection with advanced zone selection," in 31st Annual Western Protective Relay Conference, Spokane, WA, 2004.

[20] A. Guzman, J. Mooney, G. Benmouyal, N. Fischer, and B. Kasztenny, "Transmission line protection system for increasing power system requirements," in 2002 Texas A&M Relay Conference, College Station, TX, 2002.

Appendix A
Fault Analysis and MATLAB Plotting

Appendix A consists of two sections. Section A.1 illustrates the procedures of fault current calculation using analytical methods. Examples of single-line-to-ground, line-to-line, and three-phase faults are illustrated in detail. Section A.2 provides the MATLAB programming code for plotting the results of distance protection with Mho circles.

A.1 FAULT CURRENT CALCULATION

We will use Example 4.1 from Chapter 4 to illustrate the detailed procedures of fault current calculation using analytical methods. The description of Example 4.1 is shown below.

The one-line diagram of a 3-bus distribution system is shown in Figure A.1. The source voltage $V_S = 1.0$ per unit (pu). The positive- and zero-sequence source impedances are $Z_{S1} = j0.5$ pu and $Z_{S0} = j0.2$ pu. For the feeder section from Bus 0 to Bus 1, the positive- and zero-sequence impedances are $Z_{fd11} = j1.0$ pu and $Z_{fd10} = j1.5$ pu. For the feeder section from Bus 1 to Bus 2, the positive- and zero-sequence impedances are $Z_{fd21} = j3.0$ pu and $Z_{fd20} = j4.5$ pu. For each section, the negative-sequence impedance equals its positive-sequence impedance. That is to say, $Z_{S2} = Z_{S1}, Z_{fd12} = Z_{fd11}$, and $Z_{fd22} = Z_{fd21}$. The base values are $V_{llbase} = 24$ kV and S_{base} (3-phase) = 100 MVA. $S_{load1} = 3.5$ MVA at unity power factor. And $S_{load2} = 3$ MVA at unity power factor.

FAULT CURRENT CALCULATION

We will illustrate fault analysis procedures at Bus 1 as an example. The fault analysis procedures at Bus 2 can be performed similarly.

The first step is to determine the Thevenin equivalent voltage. In this example, the Thevenin equivalent voltage is 1.0 per unit.

The second step is to determine the Thevenin equivalent impedance with respect to the fault point (Bus 1). Since this system is in radial topology with a voltage source on a single end only, the Thevenin equivalent impedance equals the source impedance plus the feeder section impedance from Bus 0 to Bus 1.

Now, we start the main procedures for fault analysis using analytical methods. Pertinent theories have been illustrated in detail in typical undergraduate power system analysis textbooks [1, 2].

For single-line-to-ground (SLG) faults, we use a phase-A-to-ground fault as an example.

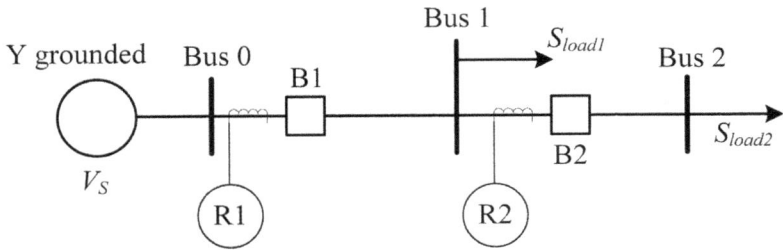

FIGURE A.1   One-line diagram for Example 4.1.

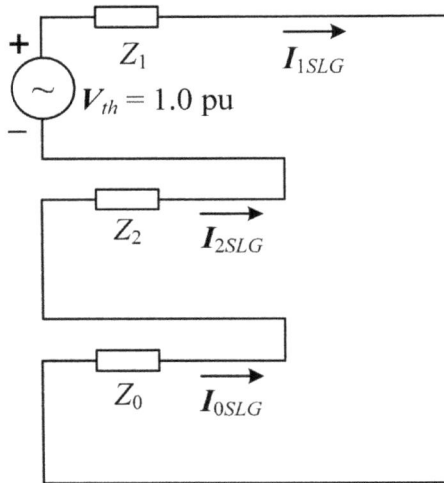

FIGURE A.2   Equivalent circuit for single-line-to-ground fault analysis.

The positive-, negative-, and zero-sequence equivalent circuits are connected in series, as shown in Figure A.2.

The Z_1, Z_2, and Z_0 in the circuit are Thevenin equivalent positive-, negative-, and zero-sequence impedances. $Z_1 = Z_{S1} + Z_{fd11}, Z_2 = Z_{S2} + Z_{fd12}$, and $Z_0 = Z_{S0} + Z_{fd10}$.

Using the equivalent circuit shown in Figure A.2, we can solve

$$I_{1SLG} = I_{2SLG} = I_{0SLG} = \frac{V_{th}}{Z_1 + Z_2 + Z_0} = \frac{1.0}{j1.5 + j1.5 + j1.7} = -0.2128j \ pu$$

$$\begin{bmatrix} I_{ASLG} \\ I_{BSLG} \\ I_{CSLG} \end{bmatrix} = A_{012} \begin{bmatrix} I_{0SLG} \\ I_{1SLG} \\ I_{2SLG} \end{bmatrix} = \begin{bmatrix} 1 & 1 & 1 \\ 1 & a^2 & a \\ 1 & a & a^2 \end{bmatrix} \begin{bmatrix} I_{0SLG} \\ I_{1SLG} \\ I_{2SLG} \end{bmatrix} = \begin{bmatrix} -0.638j \\ 0 \\ 0 \end{bmatrix} pu$$

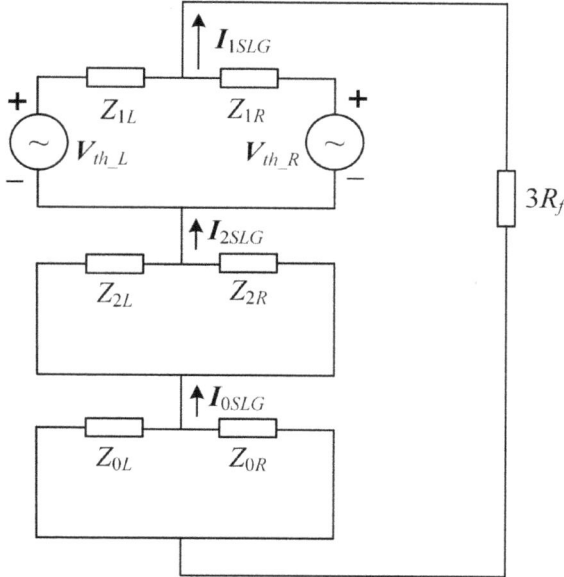

FIGURE A.3 Equivalent circuit for SLG fault analysis with two sources and fault resistance.

The parameter a in the equation is a constant, and $a = 1e^{j120 \text{ degree}}$. The base current:

$$I_b = \frac{S_{base}}{\sqrt{3}V_{llbase}} = 2405.7 \ A$$

Thus, the magnitude of single-line-to-ground (SLG) fault current is $I_{SLG\_bus1} = 1535$ A.

If the system has voltage sources on both sides and the fault event has a fault resistance R_f, the positive-, negative-, and zero-sequence circuits for SLG fault analysis should be connected in the way shown in Figure A.3.

For line-to-line (LL) faults, we use a phase-B-to-C fault as an example.

The positive- and negative-sequence equivalent circuits are connected in parallel, as shown in Figure A.4.

The Z_1 and Z_2 in the circuit are the Thevenin equivalent positive- and negative-sequence impedances. $Z_1 = Z_{S1} + Z_{fd11}, Z_2 = Z_{S2} + Z_{fd12}$. This system is in radial topology with a voltage source on a single end only. If the system has voltage sources on both ends, the Thevenin equivalent positive-, negative-, and zero-sequence impedances should consider the impedances from both sides.

Using the equivalent circuit shown in Figure A.4, we can solve

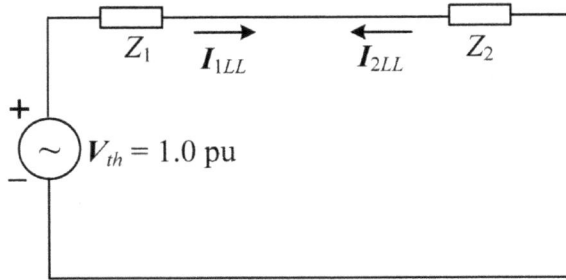

FIGURE A.4 Equivalent circuit for line-to-line fault analysis.

$$I_{1LL} = -I_{2LL} = \frac{V_{th}}{Z_1 + Z_2} = \frac{1.0}{j1.5 + j1.5} = -0.333j \; pu$$

$$I_{0LL} = 0 \; \text{pu.}$$

$$\begin{bmatrix} I_{ALL} \\ I_{BLL} \\ I_{CLL} \end{bmatrix} = A_{012} \begin{bmatrix} I_{0LL} \\ I_{1LL} \\ I_{2LL} \end{bmatrix} = \begin{bmatrix} 1 & 1 & 1 \\ 1 & a^2 & a \\ 1 & a & a^2 \end{bmatrix} \begin{bmatrix} I_{0LL} \\ I_{1LL} \\ I_{2LL} \end{bmatrix} = \begin{bmatrix} 0 \\ -0.5773 \\ 0.5773 \end{bmatrix} pu$$

The parameter a in the equation is a constant, and $a = 1e^{j120 \text{ degree}}$. The base current:

$$I_b = \frac{S_{base}}{\sqrt{3}V_{llbase}} = 2405.7 \; A$$

Thus, the magnitude of the line-to-line (LL) fault current is $I_{LL\_bus1} = 1389$ A. For a three-phase fault, only the positive-sequence impedance is involved.

$$I_{1\_3ph} = \frac{V_{th}}{Z_1} = \frac{1.0}{j1.5} = -0.667j \; pu$$

$$I_{2\_3ph} = I_{0\_3ph} = 0 \; \text{pu.}$$

Thus, the magnitude of three-phase fault current is $I_{3ph\_bus1} = 1603.8$ A.

A.2   MATLAB CODE FOR PLOTTING MHO CIRCLES

The MATLAB code for plotting the responses of ground-distance (AG, BG, and CG) elements and phase-distance (AB, BC, and CA) elements with Mho circles is shown below. Before running the code, the three-phase voltage and current phasors should be saved in vectors VAcpx, VBcpx, VCcpx, IAcpx, IBcpx, and ICcpx, respectively. After running the code, the Mho circles with impedance calculation will be plotted in two figures similar to Figure 6.3.

```
%Plot the impedance calculation with the Mho circle
   %Begin
ZAG_plot = zeros(1,len);
RAG_plot = zeros(1,len);
XAG_plot = zeros(1,len);
ZBG_plot = zeros(1,len);
RBG_plot = zeros(1,len);
XBG_plot = zeros(1,len);
ZCG_plot = zeros(1,len);
RCG_plot = zeros(1,len);
XCG_plot = zeros(1,len);
ZAB_plot = zeros(1,len);
RAB_plot = zeros(1,len);
XAB_plot = zeros(1,len);
ZBC_plot = zeros(1,len);
RBC_plot = zeros(1,len);
XBC_plot = zeros(1,len);
ZCA_plot = zeros(1,len);
RCA_plot = zeros(1,len);
XCA_plot = zeros(1,len);
%RS is the number of samples per cycle.
%The first cycle of data are set as large impedances because
%the VABCcpx and IABCcpx are zeros in the first cycle.
for i= 1:RS
    ZAG_plot(i) = 999+9999i;
    RAG_plot(i) = real(ZAG_plot(i));
    XAG_plot(i) = imag(ZAG_plot(i));

    ZBG_plot(i) = 999+9999i;
    RBG_plot(i) = real(ZBG_plot(i));
    XBG_plot(i) = imag(ZBG_plot(i));

    ZCG_plot(i) = 999+9999i;
    RCG_plot(i) = real(ZCG_plot(i));
    XCG_plot(i) = imag(ZCG_plot(i));

    ZAB_plot(i) = 999+9999i;
    RAB_plot(i) = real(ZAB_plot(i));
    XAB_plot(i) = imag(ZAB_plot(i));

    ZBC_plot(i) = 999+9999i;
    RBC_plot(i) = real(ZBC_plot(i));
    XBC_plot(i) = imag(ZBC_plot(i));

    ZCA_plot(i) = 999+9999i;
    RCA_plot(i) = real(ZCA_plot(i));
    XCA_plot(i) = imag(ZCA_plot(i));
end
```

```
for i=RS+1:len
    ZAG_plot(i) = VAcpx(i)/ (IAcpx(i)+k0*IRcpx(i)+0.0001);
    RAG_plot(i) = real(ZAG_plot(i));
    XAG_plot(i) = imag(ZAG_plot(i));

    ZBG_plot(i) = VBcpx(i)/ (IBcpx(i)+k0*IRcpx(i)+0.0001);
    RBG_plot(i) = real(ZBG_plot(i));
    XBG_plot(i) = imag(ZBG_plot(i));

    ZCG_plot(i) = VCcpx(i)/ (ICcpx(i)+k0*IRcpx(i)+0.0001);
    RCG_plot(i) = real(ZCG_plot(i));
    XCG_plot(i) = imag(ZCG_plot(i));

    ZAB_plot(i) = (VAcpx(i)-VBcpx(i))/
(IAcpx(i)-IBcpx(i)+0.0001);
    RAB_plot(i) = real(ZAB_plot(i));
    XAB_plot(i) = imag(ZAB_plot(i));

    ZBC_plot(i) = (VBcpx(i)-VCcpx(i))/
(IBcpx(i)-ICcpx(i)+0.0001);
    RBC_plot(i) = real(ZBC_plot(i));
    XBC_plot(i) = imag(ZBC_plot(i));

    ZCA_plot(i) = (VCcpx(i)-VAcpx(i))/
(ICcpx(i)-IAcpx(i)+0.0001);
    RCA_plot(i) = real(ZCA_plot(i));
    XCA_plot(i) = imag(ZCA_plot(i));
end

figure; %Plot the Mho ground distance element response
center_x = 0.5*Z1MG*cos(Z1ANG);
center_y = 0.5*Z1MG*sin(Z1ANG);
th = 0:pi/50:2*pi;
xunit = 0.5*Z1MG*cos(th) + center_x;
yunit = 0.5*Z1MG*sin(th) + center_y;
plot(xunit, yunit,'m','LineWidth',1.2);%plot the mho circle
with Zone1 reach
hold on;
x_axis(1) = 0;
y_axis(1) = 0;
x_axis(2) = Z1MAG*cos(Z1ANG);
y_axis(2) = Z1MAG*sin(Z1ANG);
plot(x_axis,y_axis,'m','LineWidth',1.2); %plot the Line
Impedance
x_axis = zeros(1,2);
y_axis = zeros(1,2);
x_axis(1) = 0;
x_axis(2) = 1.7*Z1MG;
plot(x_axis,y_axis,'k','LineWidth',0.5); %plot the x (also
called R) axis
y_axis(1) = 0;
```

```
y_axis(2) = 1.7*Z1MG;
x_axis = zeros(1,2);
plot(x_axis,y_axis,'k','LineWidth',0.5); %plot the y (also
called X) axis
hpA = plot(RAG_plot,XAG_plot,'bx');
hpB = plot(RBG_plot,XBG_plot,'r*');
hpC = plot(RCG_plot,XCG_plot,'ko');
legend([hpA hpB hpC],'AG','BG','CG');
xlabel('R (Ohm)');
ylabel('X (Ohm)');
xlim([center_x-0.7*Z1MG center_x+1.2*Z1MG]);
ylim([center_y-1.0*Z1MG center_y+1.15*Z1MG]);
title('Mho Ground Element Response');

figure; %Plot the Mho phase distance element response
center_x = 0.5*Z1MP*cos(Z1ANG);
center_y = 0.5*Z1MP*sin(Z1ANG);
th = 0:pi/50:2*pi;
xunit = 0.5*Z1MP*cos(th) + center_x;
yunit = 0.5*Z1MP*sin(th) + center_y;
plot(xunit, yunit,'m','LineWidth',1.2);%plot the mho circle
with Zone1 reach
hold on;
x_axis(1) = 0;
y_axis(1) = 0;
x_axis(2) = Z1MAG*cos(Z1ANG);
y_axis(2) = Z1MAG*sin(Z1ANG);
plot(x_axis,y_axis,'m','LineWidth',1.2); %plot the Line
Impedance
x_axis = zeros(1,2);
y_axis = zeros(1,2);
x_axis(1) = 0;
x_axis(2) = 1.7*Z1MP;
plot(x_axis,y_axis,'k','LineWidth',0.5); %plot the x (also
called R) axis
y_axis(1) = 0;
y_axis(2) = 1.7*Z1MP;
x_axis = zeros(1,2);
plot(x_axis,y_axis,'k','LineWidth',0.5); %plot the y (also
called X) axis
hpAB = plot(RAB_plot,XAB_plot,'bx');
hpBC = plot(RBC_plot,XBC_plot,'r*');
hpCA = plot(RCA _ plot,XCA _ plot,'ko');
legend([hpAB hpBC hpCA],'AB','BC','CA');
xlabel('R (Ohm)');
ylabel('X (Ohm)');
xlim([center_x-0.7*Z1MP center_x+1.2*Z1MP]);
ylim([center_y-1.0*Z1MP center_y+1.15*Z1MP]);
title('Mho Phase Element Response');
   %End
```

BIBLIOGRAPHY

[1] J. D. Glover, T. J. Overbye, M. S. Sarma, and A. B. Birchfield, *Power System Analysis & Design*, 7th Ed. Cengage Learning, 2022.

[2] A. R. Bergen and V. Vittal, *Power Systems Analysis*, 2nd Ed. Prentice Hall, 1999.

Appendix B

Transmission Line Simulation Using a Distributed-Parameter Model

When illustrating traveling wave-based protection in Chapter 10, it has been mentioned that we will only be able to capture traveling wave features if a distributed-parameter line model is used to represent a transmission line. In Appendix B, the modeling and simulation of a transmission line using a distributed-parameter line model in an electromagnetic transients simulation program is illustrated in detail.

A single-phase transmission line is modeled and simulated in the alternative transients program (ATP) [1], a type of electromagnetic transients simulation program. The circuit diagram of the simulation case is shown in Figure B.1.

The length of the transmission line is 100 km. The per-length inductance $L' = 1$ mH/km. The per-length capacitance $C' = 11.5$ nF/km. The transmission line is connected to Bus 1 with a switch. A 100-kV DC voltage source is connected to the left side of Bus 1. A terminating resistance is connected to the right side of Bus 2. Bus 1 is denoted as the sending end. Bus 2 is denoted as the receiving end. In the simulation, the switch is controlled to close at $t = 10$ milli-second.

The window for configuring transmission line parameters in the simulation program is shown in Figure B.2. A distributed-parameter line model is used in the simulation. The circuit diagram in ATP is shown in Figure B.3. The circuit diagram shown in Figure B.3 represents a short-circuit scenario in which the terminating resistance is zero. The switch at the receiving end can be opened to simulate an open-circuit scenario. In the first scenario, the transmission line resistance is set to zero. The transmission line resistance will be set as a nonzero value in later scenarios.

In the first simulation scenario, the transmission line resistance is set to zero. The terminating switch at the receiving end is set to be open, representing an open-circuit situation.

The voltage waveforms obtained at the sending and receiving ends are shown in Figure B.4. The waveform with one step is the voltage at the sending end. The waveform with an oscillating square shape is the voltage at the receiving end. It can be seen that the first step of the receiving end voltage is approximately twice the sending end voltage. A zoomed view of Figure B.4 is shown in Figure B.5 for better clarity. In Figure B.5, the letter τ represents the time difference between 10.34 ms and 10 ms, which is the time for a wave traveling from the sending end to the receiving end.

The current waveforms obtained at the sending and receiving ends are shown in Figure B.6. The waveform with an oscillating square shape is the current at the sending end. The receiving end current is flat zero.

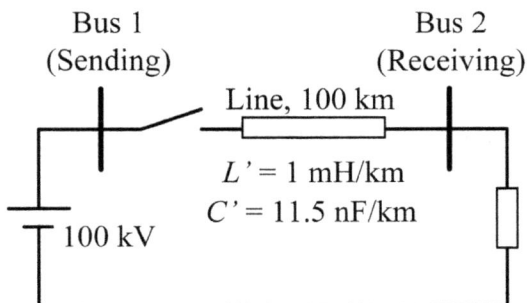

FIGURE B.1   Circuit diagram of the simulation case.

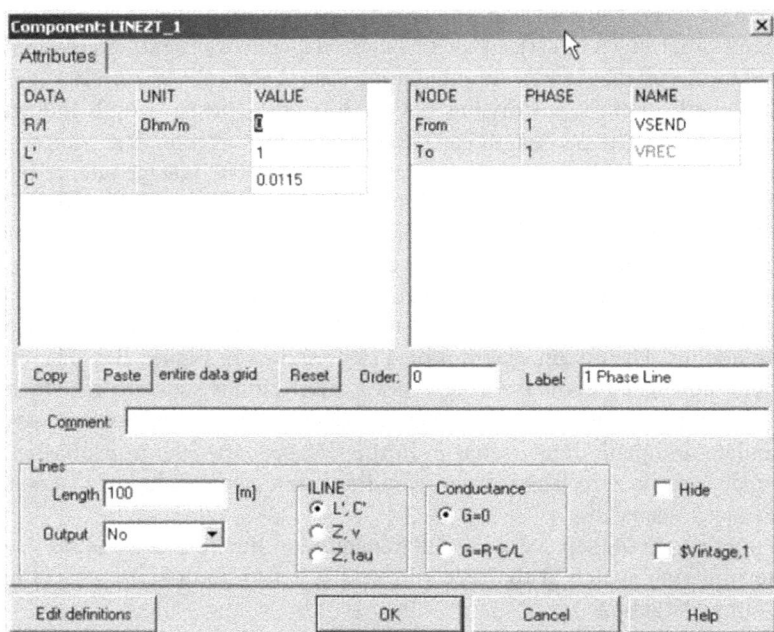

FIGURE B.2   Parameters of the transmission line.

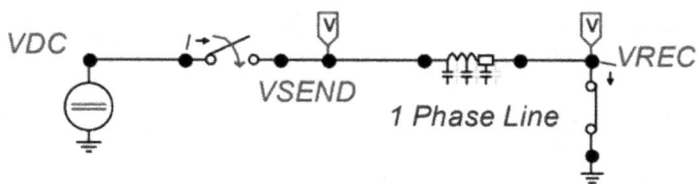

FIGURE B.3   Circuit diagram in ATP.

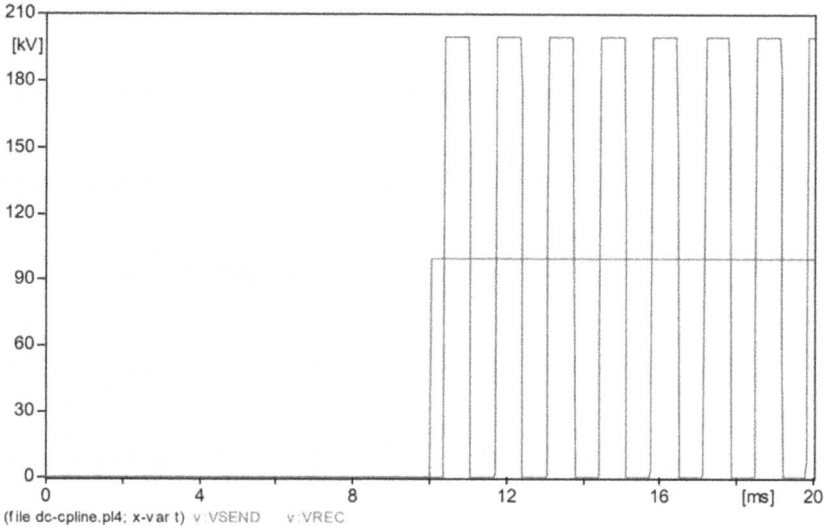

FIGURE B.4 Voltage waveforms in the open-circuit scenario with zero line resistance.

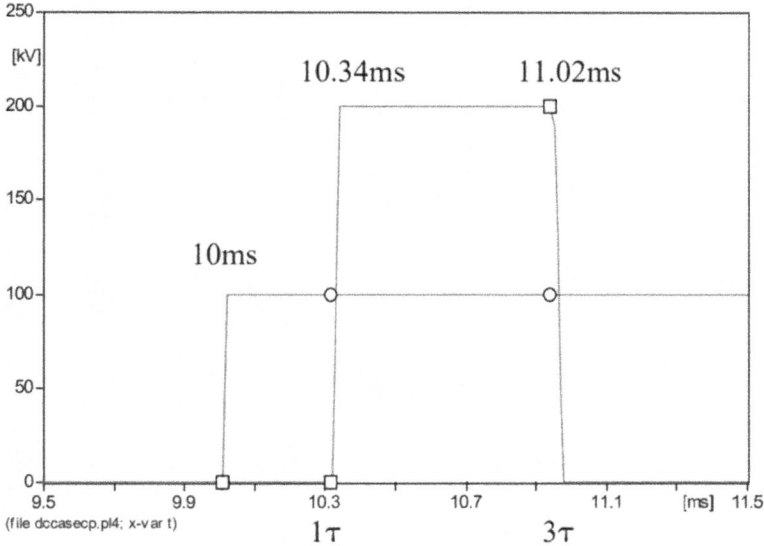

FIGURE B.5 A zoomed view of Figure B.4.

Now, we close the terminating switch at the receiving end to simulate a short-circuit scenario. The current waveforms obtained at the sending and receiving ends are shown in Figure B.7. It can be seen that the first current wave step at the receiving end is approximately twice that of the first wave step at the sending end.

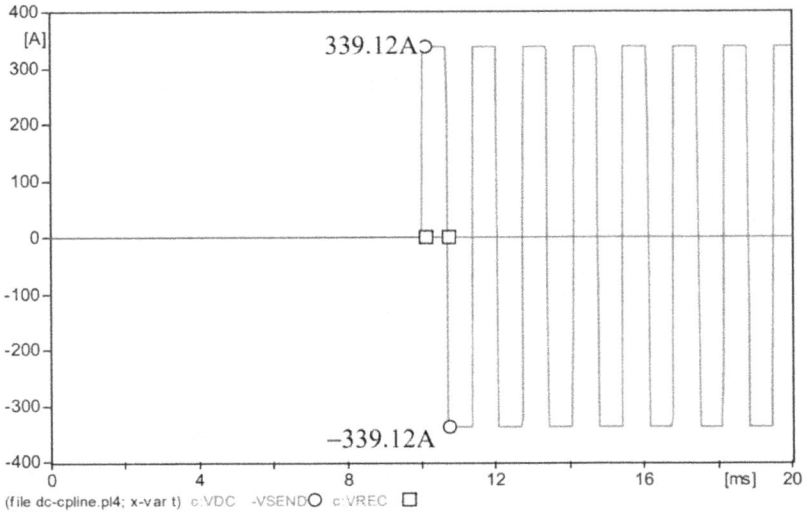

FIGURE B.6   Current waveforms in the open-circuit scenario with zero line resistance.

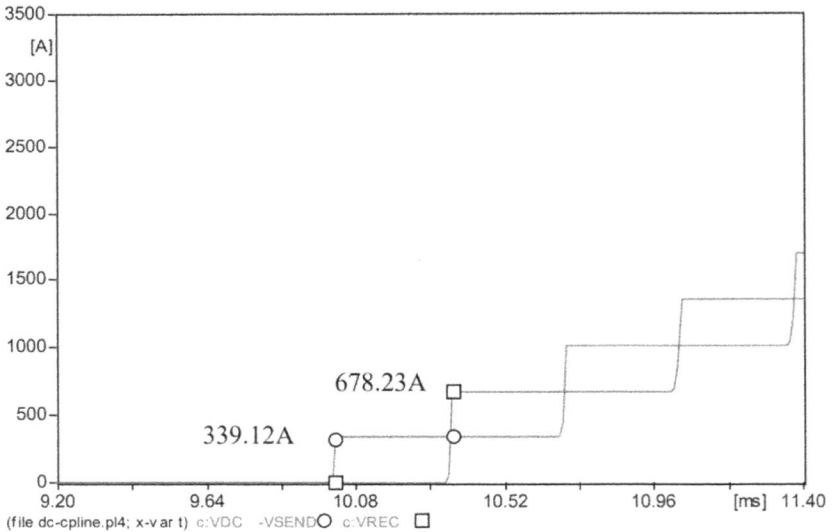

FIGURE B.7   Current waveforms in the short-circuit scenario with zero line resistance.

Now, we set the transmission line resistance to a nonzero value at $0.04\,\Omega\,/\,\mathrm{km}$. We will first simulate an open-circuit scenario and then simulate a short-circuit scenario.

In the open-circuit scenario, the voltage waveforms obtained at the sending and receiving ends are shown in Figure B.8. The waveform with one step is the

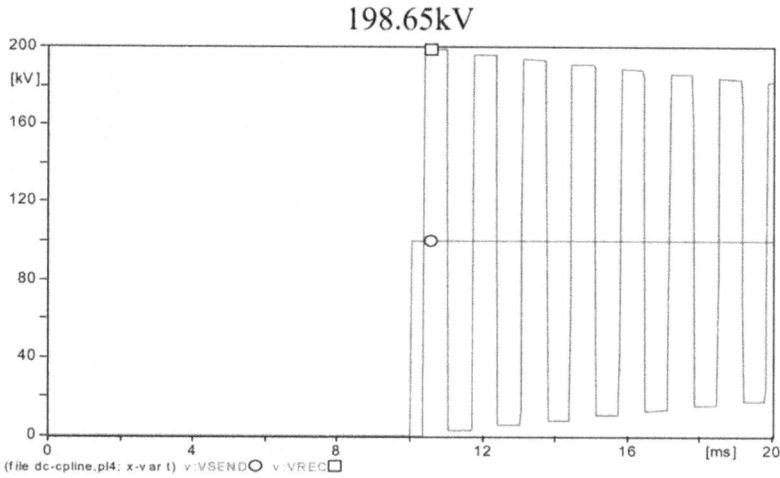

FIGURE B.8 Voltage waveforms in the open-circuit scenario with nonzero line resistance.

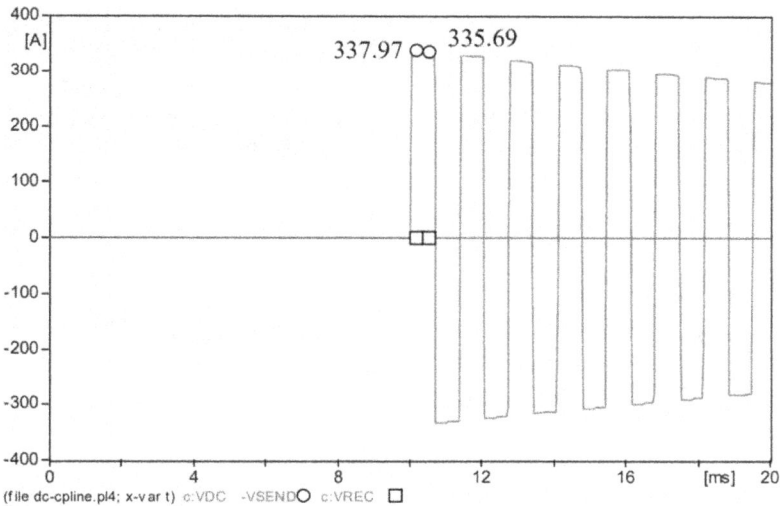

FIGURE B.9 Current waveforms in the open-circuit scenario with nonzero line resistance.

voltage at the sending end. The waveform with an oscillating square shape is the voltage at the receiving end. It can be seen that the amplitude of the receiving end voltage is decreasing over time.

The current waveforms obtained at the sending and receiving ends are shown in Figure B.9. The waveform with an oscillating square shape is the current at the

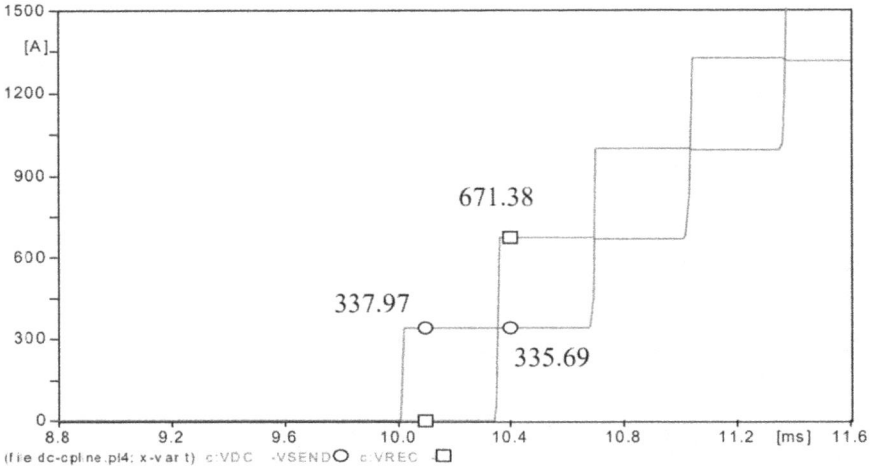

FIGURE B.10 Current waveforms in the short-circuit scenario with nonzero line resistance.

sending end. The receiving end current is flat zero. It can be seen that the amplitude of the sending end current is decreasing over time.

Now we close the terminating switch at the receiving end to simulate a short-circuit scenario. The current waveforms obtained at the sending and receiving ends are shown in Figure B.10. It can be seen that due to the transmission line resistance, the height of wave steps in Figure B.10 is less than the ones shown in Figure B.7.

BIBLIOGRAPHY

[1] A. Ametani, *Numerical Analysis of Power System Transients and Dynamics.* The Institution of Engineering and Technology, 2015.

Hints for Problems

CHAPTER 2

Problem 2.1

(a) Before the CT secondary was opened, assuming that all the secondary current flows through the resistive burden (Rb) branch. The voltage (V_1) is Isec*Rb, where Isec = 1128/CTR = 4.7 A. From the CT excitation curve, we can find the impedance Xm corresponding to V_1. The slope of the curve at V_1 is the corresponding Xm.

When CT secondary suddenly opens, the Xm (Lm) will not instantly change to a significantly different value. We may assume Xm (Lm) unchanged and all the secondary current flows to the Xm branch. We can calculate the corresponding voltage Isec*Xm, which is the initial voltage.

(b) The value of Xm (Lm) will gradually decrease due to saturation. During this process, the point shifts upward on the CT excitation curve. Finally, Xm (Lm) will stabilize at a value. With the primary side current given, we can obtain the secondary side current (4.7 A). Using the CT excitation curve data provided, we can find the corresponding voltage, which is the approximate final voltage.

Problem 2.2

Refer to the example(s) in Chapter 2.

CHAPTER 4

Problem 4.1

Refer to Appendix A.1

Problem 4.2

Refer to Example 4.1

CHAPTER 5

Problem 5.1

Check the fault type. Then, check if it has adequate amount of negative- or zero-sequence currents.

Problem 5.2

Refer to the example(s) in Chapter 5.

CHAPTER 6

Problem 6.1

Refer to the example(s) in Chapter 6.

Problem 6.2

Refer to the example(s) in Chapter 6.

CHAPTER 8

Problem 8.1

Refer to the example(s) in Chapter 6 for Mho circle plotting. You may use a program such as MATLAB or Mathcad to facilitate plotting.

Problem 8.2

Refer to Examples 8.1 and 8.2. You may use a program such as MATLAB or Mathcad to facilitate plotting.

CHAPTER 9

Problem 9.1

Find the criteria to numerically identify whether a point is in the operating region.

Problem 9.2

Refer to the illustrations and references provided in Chapters 6 and 9.

Index

Note: Page numbers in *italics* indicate a figure and page numbers in **bold** indicate a table on the corresponding page.

For Product Safety Concerns and Information please contact our EU
representative GPSR@taylorandfrancis.com
Taylor & Francis Verlag GmbH, Kaufingerstraße 24, 80331 München, Germany

www.ingramcontent.com/pod-product-compliance
Lightning Source LLC
Chambersburg PA
CBHW050500190326
41458CB00005B/1375